MARSHALL CAVENDISH

More Science Projects

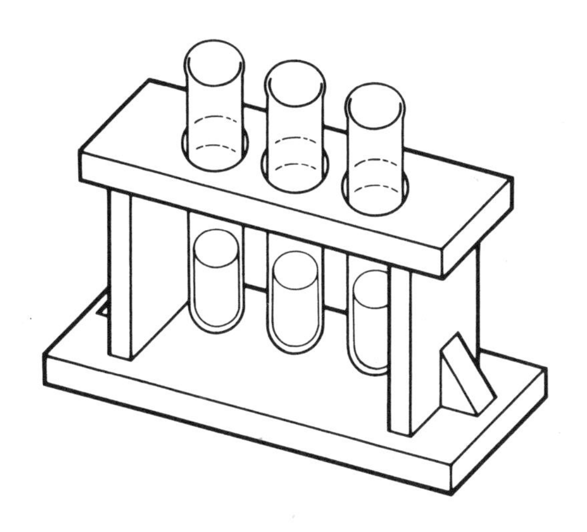

MORE SCIENCE PROJECTS

Electricity
and Magnetism

Written by Peter Lafferty

MARSHALL CAVENDISH
New York · London · Toronto · Sydney

Library Edition Published 1989

Published by Marshall Cavendish Corporation
147 West Merrick Road
Freeport, Long Island
N.Y. 11520

Printed in Italy by L.E.G.O. S.p.A., Vicenza

Library of Congress Cataloging-in-Publication Data

Lafferty, Peter.
Marshall Cavendish More Science Projects Electricity and Magnetism
written by Peter Lafferty : illustrated by Jeremy Grower, Sarah Leuton, Michael Strand.
p. cm. – (Marshall Cavendish More Science Projects II : 1)
Summary: Text and experiments provide information about the world of electricity and magnetism.
ISBN 1-85435-176-1. ISBN 1-85435-175-3 (set)
1. Electricity – Juvenile literature. 2. Magnetism – Juvenile literature. 3. Electricity – Experiments – Juvenile literature. 4. Magnetism – Experiments – Juvenile literature. [1. Electricity, 2. Electricity–Experiments. 3. Magnetism. 4. Magnetism–Experiments. 5. Experiments.] I. Grower, Jremey, ill. II. Leuton, Sarah. ill. III. Strand, Michael, ill. IV. Title. V. Series.
QC527.2.L33 1989
547–dc20

89-7152
CIP
AC

PICTURE CREDITS
Key t – Top, b – Bottom

Front cover: ZEFA

Page 6-7: E Nevill / TRH
Page 8: John Howard / Science Photo Library
Page 11: Dr Tony Brian and David Parker / Science Photo Library
Page 12: Marconi / TRH
Page 13: Canon / TRH
Page 18 t: Ann Ronan Picture Library
Page 18 b: Ann Ronan Picture Library
Page 19: The Fotomas Index
Page 21: BBC Hulton Picture Library
Page 23: Dennis Di Cicco / Science Photo Library
Page 26-27: Royal Institution
Page 33: The Mansell Collection
Page 36-37: Paul Brierley
Page 43: GPT Consumer Products Ltd

Artwork by: Tony Gib[illegible] / Bernard Thornton Artists
Jere[illegible]er / B L Kearley Ltd
Sarah Leuton / B L Kearley Ltd
Michael Strand / B L Kearley Ltd

CONTENTS

GOLDEN RULES

This book contains lots of scientific facts, experiments, and projects to help you find out more about electricity and magnetism. Whenever you try one of the experiments, make sure you read all about it before you start. You'll find a list of all the things you need, a step-by-step account of what to do, and finally an explanation of why and how your experiment works.

▶ Always watch what happens very carefully when you're doing an experiment, and if you find it doesn't work the first time, *don't* give up. Consider what could have gone wrong and then read through the experiment once more. Check that everything is just right, and then try, try again. Real scientists may have to do an experiment several times before they get a worthwhile result.

▶ Because you will be such an active scientist, it's a good idea to start collecting for your laboratory. Nearly everything you need for the experiments can be found around your house. For example, bottles, cans, and pieces of cardboard and paper will often be used, so when you see your parents throwing away

GOOD SCIENTISTS...

ALWAYS THINK SAFETY FIRST

Famous scientists take precautions to avoid danger, so that they live to see their results and enjoy their fame. In any project or experiment, especially one you have thought up yourself, consider what it is you are trying to show and have a good idea of what should happen. Don't do any experiment "just to see what happens". Always plan carefully.

ALWAYS KEEP A NOTEBOOK

Whenever you are involved in scientific activity, keep a *Science Notebook* by your side and fill it with notes and sketches as you go along. Get into the habit of writing up your experiments and observations – your notes will be useful in the future.

ALWAYS FOLLOW GOOD ADVICE

Advice and instructions, like the leaflets that come with pieces of equipment, should be read and understood. They are there for your safety and help. Good scientists think for themselves, but they are also wise and listen to what others have to say.

useful containers, offer to wash them and add them to your collection. General things like rulers, spoons for measuring, thumb tacks, and scissors will also come in handy. You'll also need colored pens and paper for lots of the experiments, as well as tape and glue. Finally, you'll need a worktop for your experiments; if possible, it should be near a sink. Store your collection in a nearby cupboard or cardboard box.

▶ Always tell your parents what you are doing. Sometimes you'll need their help. And when it comes to special equipment like batteries or chemicals, they'll know where to get them. Your parents may also help you build wooden stands or nail things down when needed. And if you need to use matches, cut things out, or drill holes, **always** ask their permission first.

▶ Good scientists are clean and tidy! They always remember to clear up when they have finished their experiments. So after you have completed your project, throw away anything you won't need again and clean everything else, ready for next time.

NEVER PLAY WITH ELECTRICITY

Don't play with electricity or transformers. All the experiments in this book will work with a small battery. Batteries with screw-on terminals are easiest to use. You can use batteries with spring terminals, but you will have to wind the wires onto the terminals.

NEVER PLAY WITH CHEMICALS

Avoid mixing chemicals and powders unless you are sure that you know what is going to happen, and always use small quantities. Dangerous chemical mixtures can explode or start a fire or burn your eyes and skin. Make sure any chemicals you keep are properly stored in jars and are correctly labeled.

NEVER FOOL WITH HIGH-PRESSURE EQUIPMENT

Do not play around with gas or liquids under pressure, especially in containers like aerosol cans – even if they seem empty. They can blow up in your face. Dispose of empty aerosols carefully, and *never* put them in or near a fire.

ELECTRIC POWER

Electricity is the servant of the modern world, helping us in many different ways. It can heat and light our homes, make music to entertain us, and cook our food. In factories, it lifts heavy weights, moves trucks, and powers the machines that make and pack goods. Weather forecasters and scientists use electricity to do complicated calculations on their computers. Electric trains carry passengers swiftly across the country. Even gasoline-driven cars need electricity from their batteries in order to run. A huge city hums with life all night long thanks to electricity.

Electricity is also a messenger. When you use the telephone to talk to a friend, your voice is carried along the telephone wires by electricity. It carries messages to all corners of the world at great speed. Radio and television signals are carried through air and space by ripples, or waves, of electricity and magnetism. Light is also made up of these waves. So is the life-giving heat we get from the sun.

Electricity is our most useful source of energy and power. It can be sent long distances along wires, which means that power stations that make electricity can be far away from the cities. Electricity can also be changed into other forms of energy, such as heat, sound, and light. Electricity is clean – there is no smoke or ash from an electric heater. Small batteries produce electricity for flashlights and radios that can be carried around.

We even use electricity in our bodies. Many of the body's processes involve electricity; for example, when your heart beats, it produces tiny pulses

A city never sleeps. People go to films, shows, and restaurants in the evenings. All through the night, the street lights and lights of the skyscrapers shine, and traffic moves through the streets. Without electricity, all this activity would stop at sunset.

of electricity. When you dream at night, your brain is making electricity. Electrical messages from your brain make your muscles move. Some animals even use electricity to catch food. The electric eel and the electric ray, for example, use electricity to stun their prey.

WHAT IS ELECTRICITY?

About 2,500 years ago, a Greek scientist called Thales made a discovery. He rubbed a piece of amber (fossilized resin) with a silk cloth. He noticed that the amber could then pick up small pieces of straw or feathers. He had discovered electricity.

Sometimes you can see sparks of electricity when you comb your hair in a dark room. And sometimes there are sparks when you take off a nylon shirt or blouse. The same thing happens when a storm cloud gets an ***electric charge***. The electric charge is made when drops of water in the clouds rub together. Sparks fly between clouds or to the ground. These sparks of electricity are called lightning.

This girl has been given an electrical charge by her teacher. Her hair is standing on end because each hair is repelled by the electricity on the others. If you comb your hair with a plastic comb, the comb will then pick up small pieces of paper, as it is filled with electricity.

SCIENCE DISCOVERY

High flier!

Benjamin Franklin was a famous American inventor and scientist. In 1752, he did a dangerous experiment by flying a kite during a thunder and lightning storm. Electricity flowed down the string of the kite, making a small spark on the metal key at the bottom of the string near his hand. This showed that a lightning bolt was really just a large electric spark. After the experiment, Franklin invented the lightning conductor. This is a metal strip that runs from the top of a building to the ground. It carries away the electricity if lightning strikes the building.

Electrons – pieces of electricity

Electric charges are caused by ***electrons***. These are tiny bits of negative electricity, which are in the atoms that make up all matter. Atoms, which are too small to be seen, are like tiny building blocks that join together to make bigger objects. The smallest speck of dust contains millions of atoms.

Inside each atom are electrons. When you pull a comb through your hair, you disturb the electrons in the atoms of the comb, and some of the electrons from the comb are left behind on your hair. The result is that the hair gets a negative charge because of its extra electrons. The comb gets the opposite charge, because it has lost some electrons.

If some electrons jump from your hair to the comb, they make a spark as they travel through the air. Electrons can also travel along wires. When they do, they make an ***electric current*** along the wire. Electricity that stays in one spot is called ***static electricity***.

Positive and negative electricity

Scientists have discovered that there are two kinds of electricity, positive and negative. Electric charges only attract each other if they are of opposite kinds, so a negative charge attracts a positive one. When you pull a comb through your hair, the comb gets a positive charge, and your hair gets a negative charge, so the hair is attracted to the comb. Two charges that are the same repel each other.

SCIENCE PROJECT

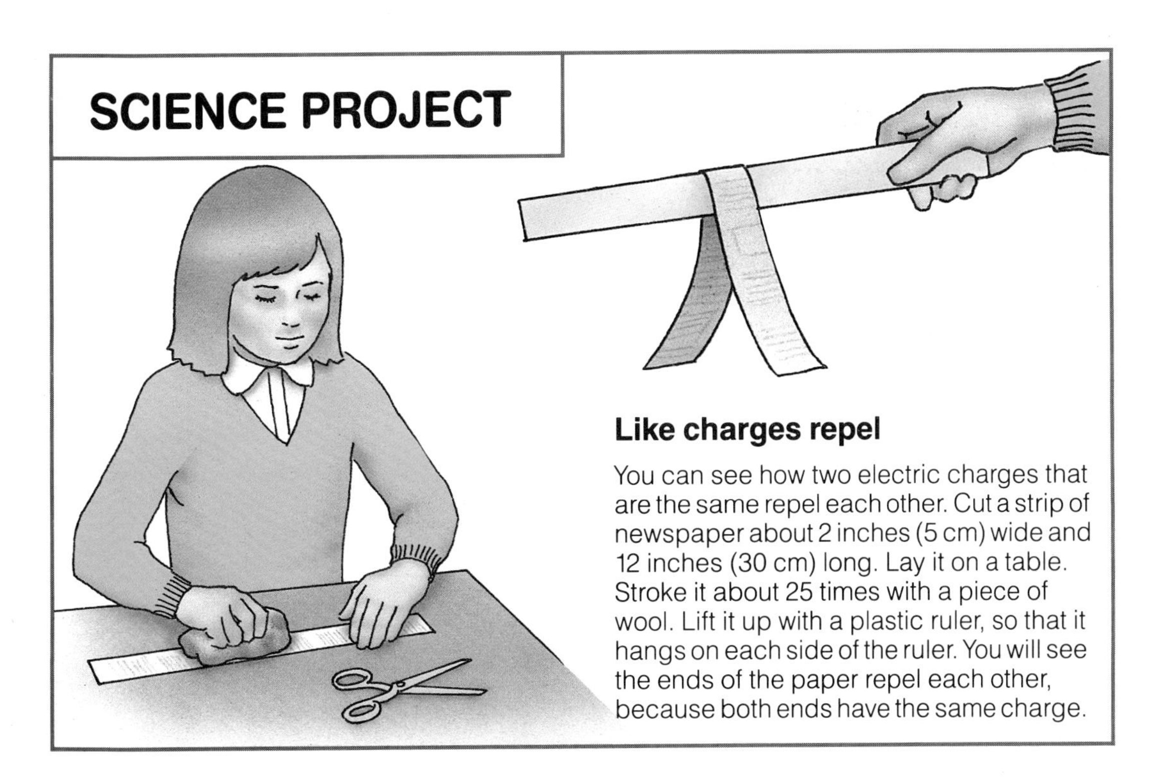

Like charges repel

You can see how two electric charges that are the same repel each other. Cut a strip of newspaper about 2 inches (5 cm) wide and 12 inches (30 cm) long. Lay it on a table. Stroke it about 25 times with a piece of wool. Lift it up with a plastic ruler, so that it hangs on each side of the ruler. You will see the ends of the paper repel each other, because both ends have the same charge.

ELECTRONS ON THE MOVE

If you connect the ends of a ***battery*** to a small light bulb using pieces of wire, you will make a simple ***electric circuit***. The bulb lights up because a current is flowing along the wires. The battery is like an electron pump, pushing electrons along the wires and through the bulb. Electrons are very small, and it takes billions to make an electric current.

Electrical pressure

In some ways, an electric current flowing along a wire is like the flow of water in a pipe. In both cases, there must be something to push the flow along. A pump pushes water along a pipe, and a battery is needed to push electrons along a wire. The push or pressure of the battery is called the ***electromotive force***, or ***potential difference***. This pressure is measured with a ***voltmeter*** in units called ***volts***. The higher the voltage of a battery, the stronger the electrical pressure it provides.

When we do experiments with electricity, we use small batteries of 3, 4½, or 6 volts. This is not dangerous. Never use electricity from outlets for experiments; its voltage is 120, high enough to kill you.

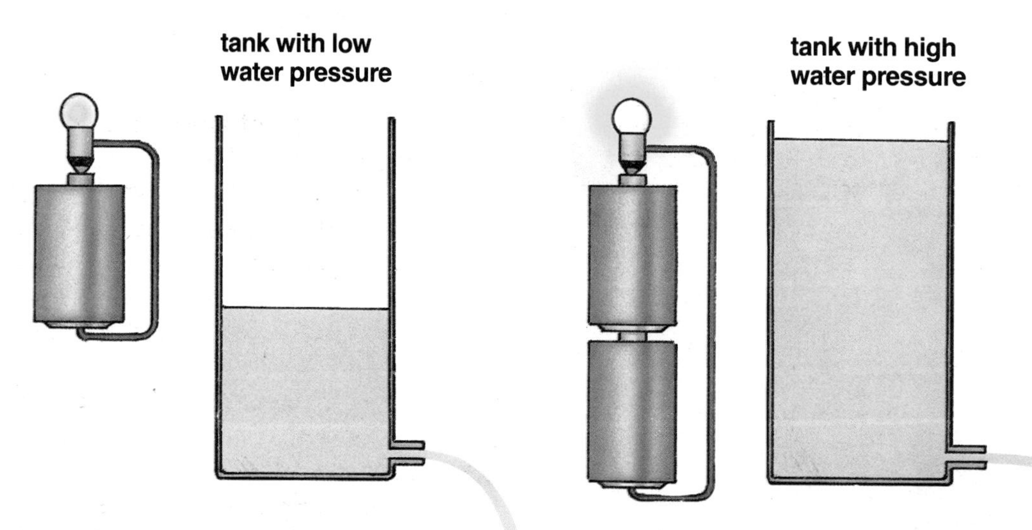

Here, the left-hand picture shows a single battery dimly lighting a bulb. It is like the water container with a small amount of water. The flow of water is weak because there is not much pressure.

In the right-hand picture, there are two batteries and the bulb is twice as bright. The water container holds twice as much water, and the flow of water is twice as strong.

Conductors

The metal wires used in an electric circuit contain electrons that can move. When a battery is connected to the ends of the wire, the electrons all move along in the same direction, and an electric current flows. Metals are called *conductors* if electricity can flow in them.

When electrons flow along a metal wire, they bump into the atoms of the metal. The atoms resist the flow of the electrons. This resistance is like the friction between the sides of a pipe and the water flowing through it.

An electric light bulb works through resistance. When the electrons reach the bulb at the end of the wire, they have to pass through a thin wire, called a filament. The filament is so thin that collisions between electrons and atoms in the wire are very frequent.

In a good conductor such as a copper wire, the electrons move in all directions (A). When the wire is connected to a battery and becomes part of an electric circuit, the electrons move in one direction only, and a current flows (B).

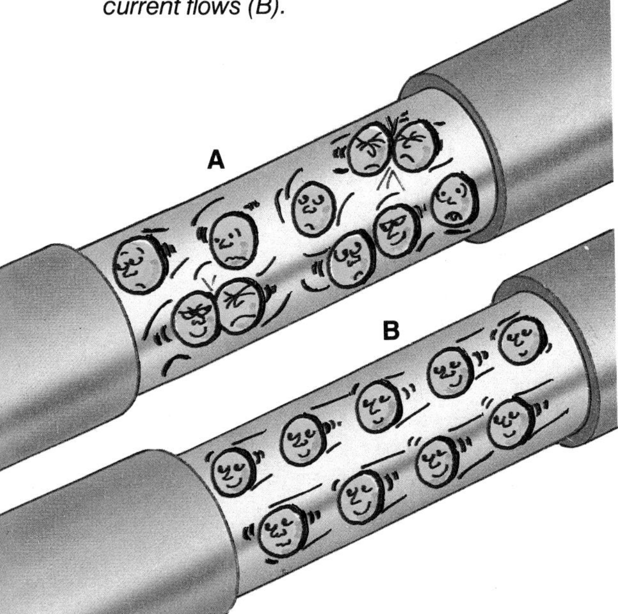

This picture of a woodworm beetle was taken with an electron microscope. Electron microscopes shine a beam of electrons instead of light onto an object. The picture of the object, magnified up to a million times, is shown on a screen like a television.

These collisions increase the temperature of the filament and it becomes hot and glows, giving off light.

Insulators

Insulators do not let electricity flow. In an insulator, the electrons cannot move about, but are firmly attached to the atoms. Insulators protect us from electricity. If you could touch the metal in the wires linked to the household electricity supply, you would get a nasty shock. For this reason, the metal wires are covered with a plastic coating. The plastic is an insulator and stops electricity from escaping from the wire. Electrical appliances and tools also have handles made of insulators, like rubber and plastic.

Never touch anything electrical if you have wet hands. Water conducts electricity, and you would get a dangerous shock.

ON THE CIRCUIT

At Christmas, many people have colored lights on their Christmas tree. In the past, the bulbs for these lights were connected in a line, and the bulbs at each end of the line were connected to the battery or power supply. This arrangement is called a *series circuit*. The electric current flows through all the bulbs, one after the other. If one bulb fails, the circuit is broken and all the lights go out.

Fortunately, modern lights are usually connected in a different way, with each bulb connected to the battery. This arrangement, known as a *parallel circuit*, has a big advantage over a series circuit because even if one bulb fails, the others still shine.

These two bulbs are connected side by side. They are said to be in parallel. Both bulbs receive the full voltage of the battery and they both glow brightly, but the battery will only last a short time.

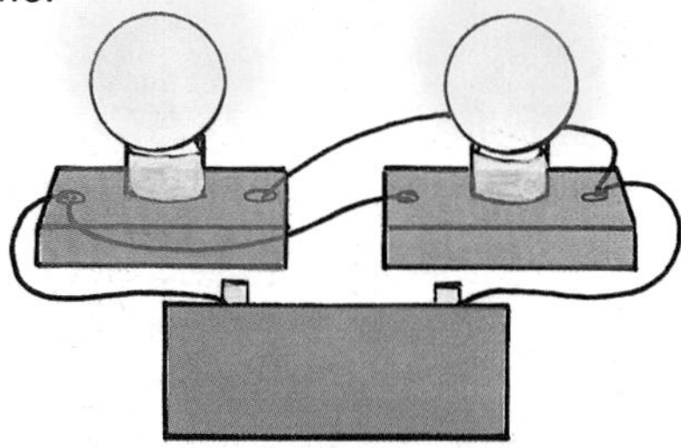

These two bulbs are connected in a line. They are said to be in series. The voltage is shared between the two bulbs.

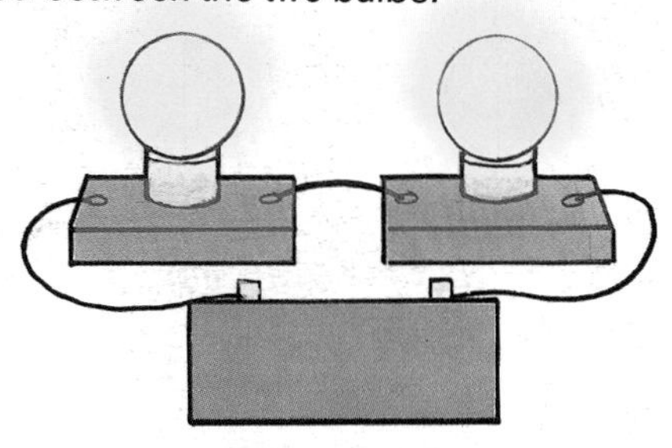

Inside a radio

Series and parallel circuits are the simplest types of electric circuits. All other circuits are just combinations of these two. If you look inside a radio,

SCIENCE IN ACTION

Making chips!

This engineer is designing the electrical circuits for a new integrated circuit, which are complete circuits formed on a tiny silicon chip. She is using a computer to help her. In front of her is a large piece of photographic film showing the circuit. When the circuit is correct, the film will be reduced in size until it is about ¼ inch (5 mm) square. It will then be photographically copied onto a small piece, or chip, of very pure silicon. The chip is then baked in an oven. This process makes it possible for an electric current to pass along the circuit lines. Then the chip is put into a small plastic box, with metal legs at the sides. The chip is connected by tiny gold wires to the legs on the case.

for example, you will see what seems like a mass of wires. This is the electric circuit that makes the radio work. It seems complicated, but in fact, it is simple – the components are connected end to end in series, or side by side in parallel.

There are several types of electrical components in a radio, including resistors, which are used to control electric currents; capacitors, which can store electricity; and transistors, which act like switches and amplify, or increase, a current.

The circuit board

In a modern radio, you will also see integrated circuits, which look like small plastic boxes with metal legs on each side. Inside the box is a small piece, or chip, of a material called silicon. This has a complete electric circuit built on it, and all the parts of the circuit, such as the resistors, transistors and the wires connecting them, are built into the tiny chip. Integrated circuits are used extensively in radios, television sets, automatic cameras, computers, cars, and microwave ovens.

Left:
An integrated circuit is a complete electrical circuit formed on a tiny chip of silicon. The silicon chip is contained in a package to make it easier to handle. Silicon chips are so small that they can even pass through the eye of a needle.

Below:
A modern camera contains many integrated circuits. They are used to focus the camera automatically and to control the amount of light entering the lens.

PROJECT 1

Electric quiz board

Make a quiz board that lights up when you answer questions correctly. You need a piece of stiff white cardboard 12 x 24 inches (30 cm x 60 cm), a pen, 10 paper fasteners, 8 lengths of insulated wire about 12 inches (30 cm) long, a 4½ volt battery, a small flashlight bulb suitable for a 4½ volt battery, and a bulb holder.

STEP 1 Think up five tricky questions. Write the questions in a column down the left-hand side of the cardboard. Write the answers in a column down the right-hand side of the cardboard. Mix them up so that the questions are not opposite their answers. With a sharp pencil, make a hole through the cardboard next to each question and answer.

ELECTRIC QUIZ BOARD

QUESTIONS

What is the Capital of France? o

Who is the Prime Minister of England? o

What is 100 divided by 4? o

How many is a dozen? o

What is paper made from? o

ANSWERS

o Mrs Thatcher

o 12

o Wood

o Paris

o 25

STEP 2 Push a paper fastener through each hole so that the "legs" are at the back of the board. Bend back the legs of the fastener.

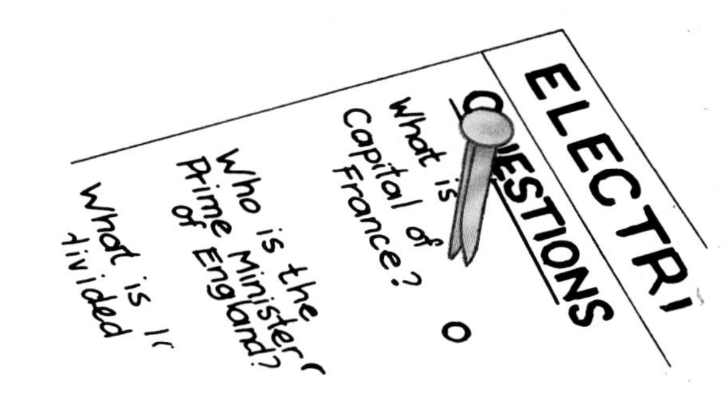

STEP 3 Wind one end of a piece of wire around the legs of a paper fastener near a question. Wind the other end of the wire around the legs of the paper fastener near the answer to the question. Do this for every question. You will end up with a wire going from each question to its correct answer.

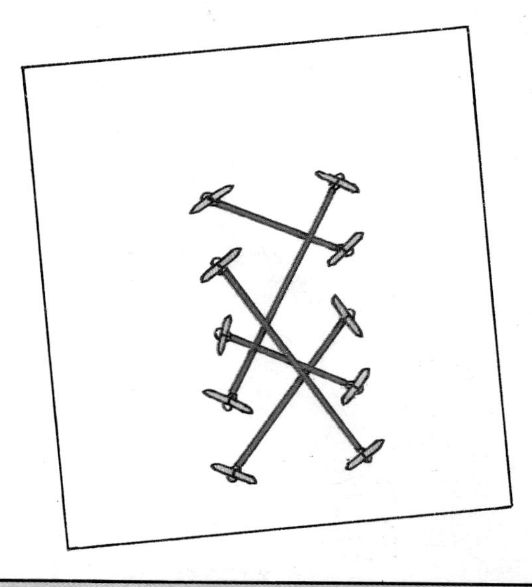

STEP 4 Wind a piece of wire around each terminal of the battery. Wind the other end of one of the wires to a connector on the bulb holder. Connect another piece of wire to the other connector on the bulb holder. Screw the bulb into the holder. When you touch the free ends of wire together, the bulb should light up.

STEP 5 Ask a friend to choose a question and touch the fastener next to it with the wire from the battery. Then ask her to touch the fastener beside the correct answer with the wire from the bulb. If she's right, the bulb will light up. If not, she – or your wiring – is wrong! Touching the correct fastener should complete the circuit with the wires behind the board.

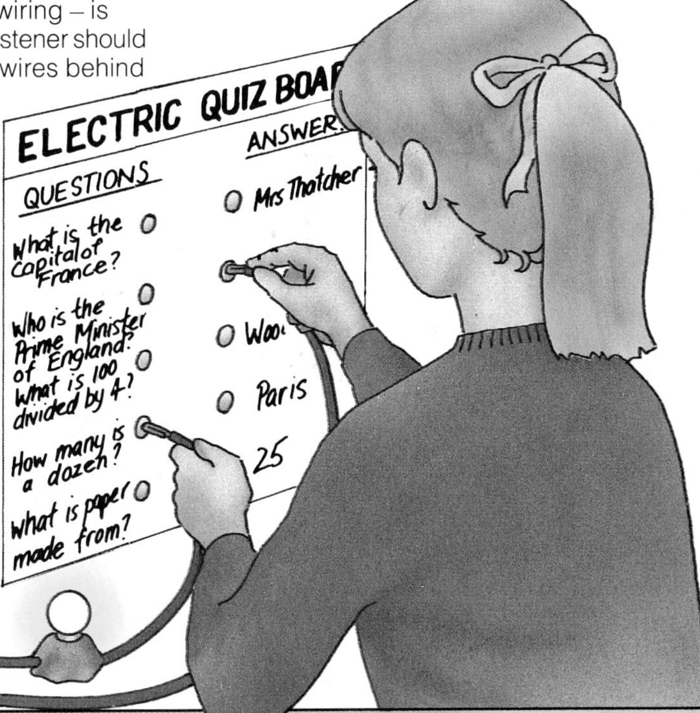

PROJECT 2

Test your skill

This game tests whether you have steady hands. You will need 3 pieces of thin insulated wire about 8 inches (20 cm) long, a 4½ volt battery, a bulb holder, a small flashlight bulb suitable for a 4½ volt battery, a piece of thick, stiff wire about 12 inches (30 cm) long, a piece of thin, stiff wire about 8 inches (20 cm) long, and some modeling clay.

STEP 1

Wind a piece of the insulated wire around each terminal of the battery. Wind the other end of one of these wires to a connector on the bulb holder. Connect the third piece of wire to the other connector on the bulb. Screw the bulb into the holder. When you touch the free ends of wire together, the bulb should light up.

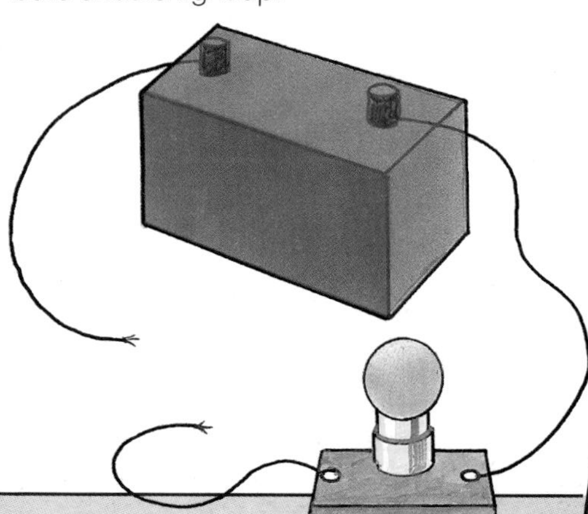

STEP 2

Bend the thick wire into a curvy, up and down, shape. Wind the end of the wire from the battery around one end of the thick wire.

Quick bulb holder

If you have no bulb holder, make one from modeling clay. Use a small piece of clay as a base. Press the bulb into the clay, and press one wire into this base so that the wire touches the bottom of the bulb. Twist the other wire around the shaft of the bulb.

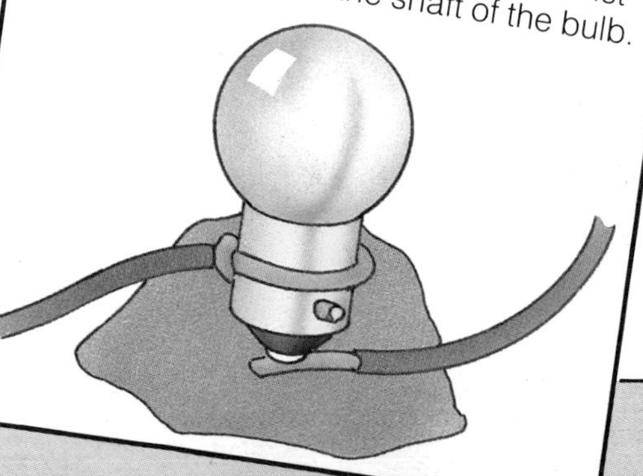

STEP 3

Bend one end of the thin, stiff wire into a small loop, and wind the wire from the bulb holder around the other end. Thread the thick wire through the loop. Finally, stand the thick wire upright using the clay.

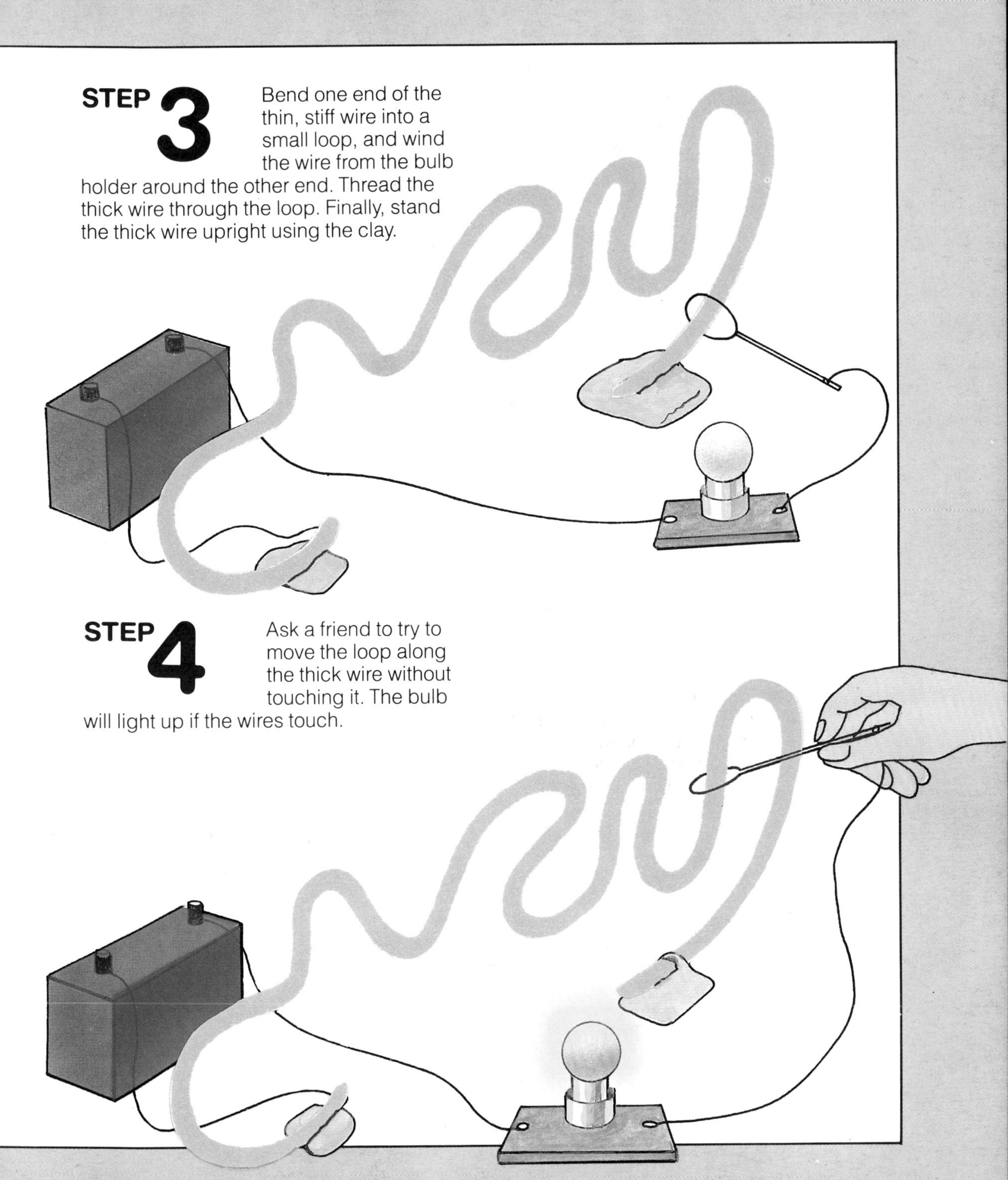

STEP 4

Ask a friend to try to move the loop along the thick wire without touching it. The bulb will light up if the wires touch.

MAGIC MAGNETS

Have you seen a piece of iron pick up a small pin or piece of metal? This strange power is called magnetism. A ***magnet*** is any substance – for example, iron or nickel – that attracts iron. It also points north-south when it is suspended. Magnets were discovered over 2,000 years ago.

Some people think that a shepherd named Magnes, who lived on the island of Crete, discovered the first magnet. One day, Magnes poked a long walking stick with a metal ring on the end into a stream. To his surprise, pieces of rock stuck to the metal ring. These rocks were natural magnets. They could pick up small pieces of iron, such as horseshoe nails.

Others think that magnetic stones were first found in Greece, near a place called Magnesia. No one knows if these stories are true, but such stones became known as magnets.

The Chinese also found magnets long ago. At first, they thought they were magic and used them to tell fortunes. The fortune teller spun a piece of magnetic rock shaped like a spoon. Soon it was noticed that the handle of the spoon always pointed south when it stopped spinning.

Sailors used these magnetic spoons to help find their way in foggy or cloudy weather. The stone that the spoons were made from was called "leading stone" or lodestone. Later, the Chinese made long, thin magnets which floated on water and turned toward the north. They were called compasses.

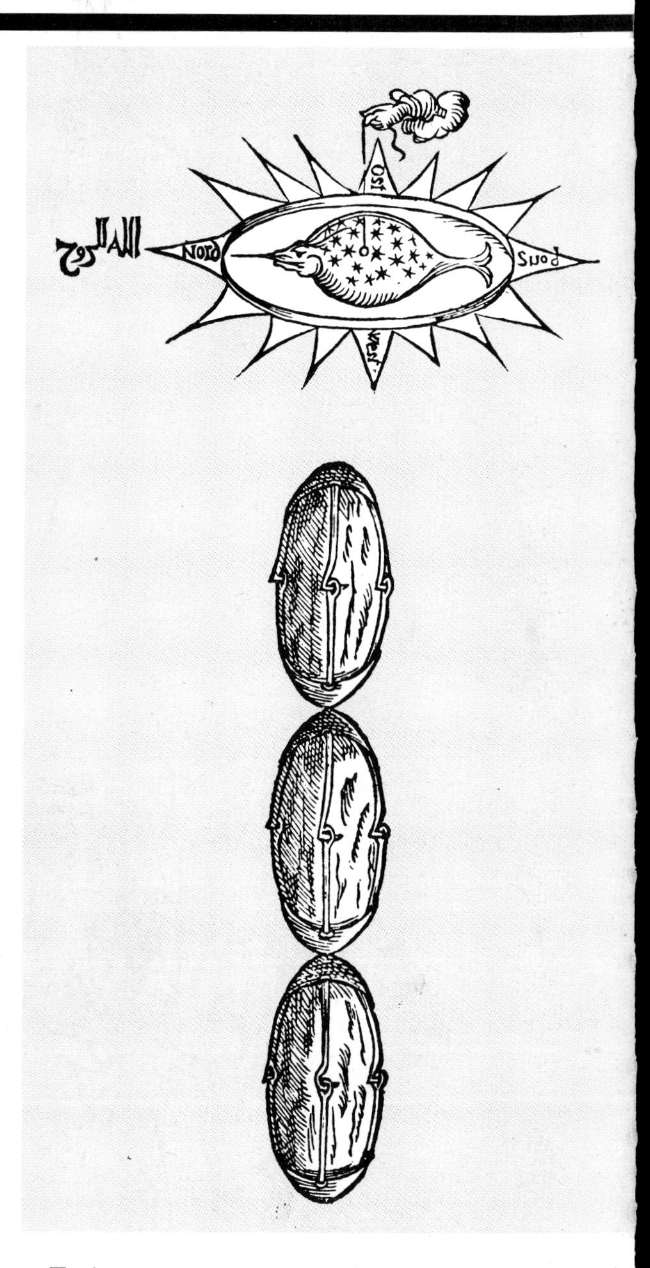

Today, we use magnets in many ways. Did you know that there are magnets in your television set? There is also one in your refrigerator door that helps

Lodestones, like the two pictured on the left dating from the early seventeenth century, were the earliest kind of compass. They gave rise to the magnetic compass, an early form of which is pictured on the right.

keep the door closed. They are in your telephone and record player, too. The tapes used in a tape recorder have a magnetic coating, and computers remember information by writing it onto a magnetic disk. Huge magnets are used in electricity power stations to help generate electricity.

MAGNETIC ATTRACTION

The magnets we use today are made from iron and steel. Some types of steel make very strong magnets that keep their power for a long time. Some magnets, called bar magnets, are shaped like chocolate bars. Others are shaped like horseshoes.

Magnets pull some things toward them. Metals like steel, iron, cobalt, and nickel are all attracted to a magnet. Copper, plastics, rubber, paper, hair, and glass are not attracted to magnets. They are said to be non-magnetic.

How many things can you find that are attracted to a magnet? What do they all have in common? Try a pin, bottle, wood, paper clip, iron wire, copper wire, different types of coins, and aluminum foil.

Powerful poles

If you have a magnet, it is fun to play with iron filings. These are shavings of iron about the size of sand grains. If a magnet is dipped into the filings, they will stick to the magnet. But they will stick most strongly at two points, called the ***magnetic poles***, where the magnet's power is strongest.

The two poles are called the north and south poles. If a magnet is hung on a thread so that it can swing freely, its north pole will point to the north and its south pole will point to the south. You can find out why on page 22.

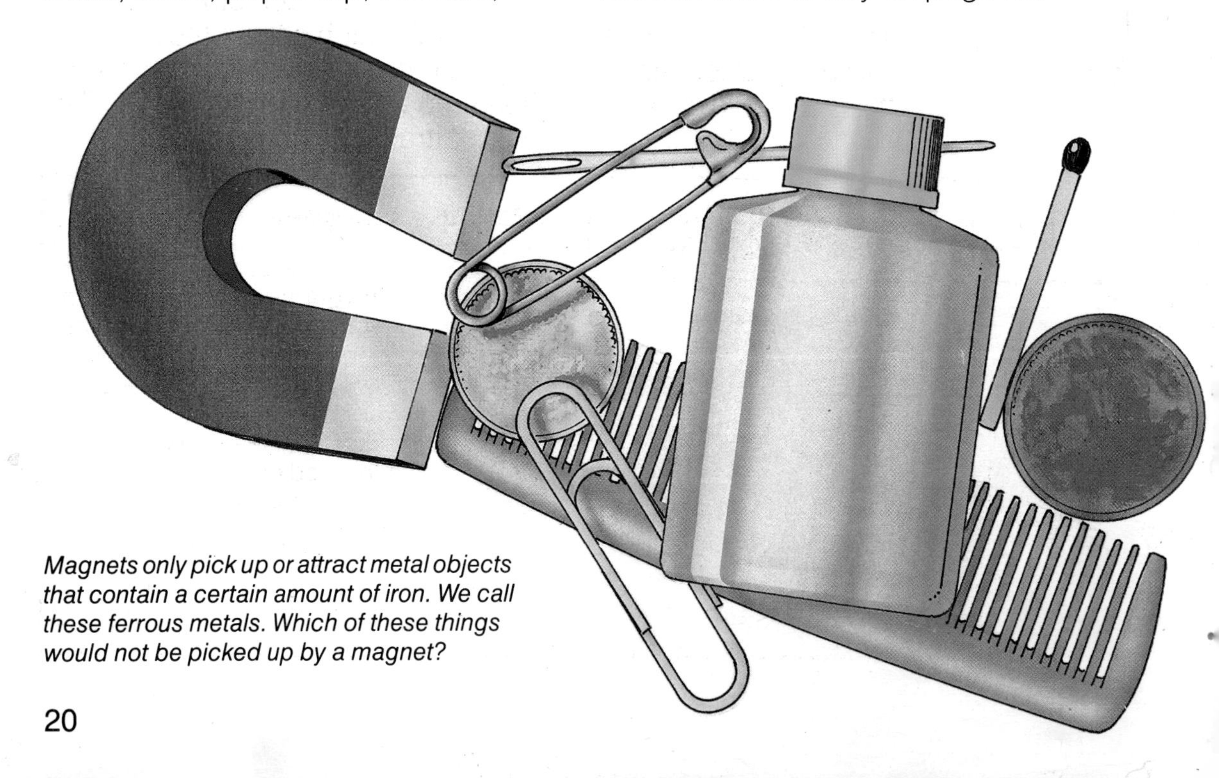

Magnets only pick up or attract metal objects that contain a certain amount of iron. We call these ferrous metals. Which of these things would not be picked up by a magnet?

SCIENCE DISCOVERY

Magnet man!

William Gilbert was an Englishman, born in 1540. He was Queen Elizabeth I's doctor. In 1600, he wrote a book called *De Magnete*, in which he told how steel rods could be made into magnets by stroking them with lodestone. He also described how a steel rod became a magnet when it was pointed northward and hammered. He believed that the earth was a giant magnet. To test this idea, he made a ball of lodestone and put many small iron needles on its surface. All the needles pointed north–south, just like magnets on the surface of the earth. William Gilbert was also interested in electricity. He was also the first person to use the word "electricity".

William Gilbert explaining his discoveries to Queen Elizabeth.

Poles apart

If you have two magnets, you will notice that they do not always attract each other. If you put the north pole of each magnet together, the magnets will push apart, or repel, each other. Two south poles also repel each other. Magnets only attract each other if their opposite poles are close together. Remember, "Like poles repel; unlike poles attract."

SCIENCE PROJECT

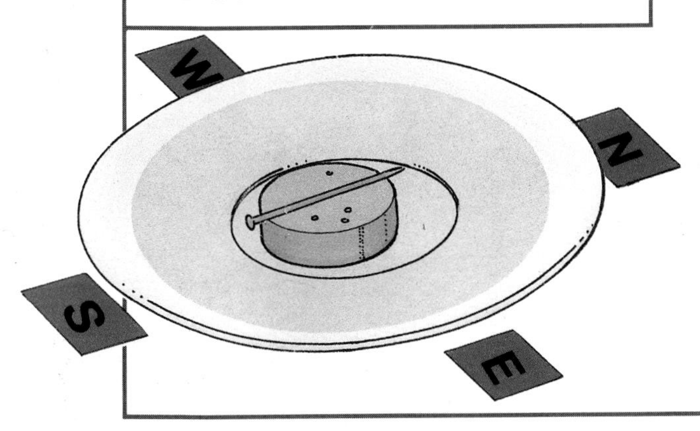

Making a compass

You can make a simple compass using a piece of thin, straight iron wire or a pin, a flat cork, and a saucer of water. First, magnetize the pin by stroking it about 50 times with one end of a magnet. Next, place the cork in the saucer of water and make sure it is floating in the middle of the saucer. Gently place the magnetized pin onto the cork. The cork and pin will turn, and the pin will point north–south. Put pieces of paper marked "North," "South," "East," and "West" around the saucer. You need to know roughly which direction is north.

FORCE FIELDS

When you hold two magnets near each other, you can feel the effect. The magnets will either attract each other or push each other apart. They do not have to be touching each other to do this. You can feel the effect of a magnet in the space around it. Scientists describe this by saying that there is a *magnetic field* around the magnet. A magnetic field is the space around the magnet where you can feel its force.

A pattern of imaginary lines, called *lines of force*, can be used to describe the magnetic field. These lines run from one pole of the magnet to the other pole. They are closest together near the poles where the effect of the magnet is strong. Away from the poles, the magnetic effect is weaker, and the lines of force farther apart.

Earth's magnetic field

The earth is surrounded by a magnetic field. It acts as a giant magnet. Its magnetic field affects compass needles, making the north pole of a compass needle point to the north. The magnetic field of the earth is produced by the molten metal which is found deep within the earth's core. As the earth spins, electric currents are created in the molten metal, which produces the magnetic field.

The earth's magnetic field has other effects. It traps small electrical particles which reach the earth from space. These particles form belts in space high above the earth. Spacecraft carrying astronauts keep away from these belts because the particles are

SCIENCE PROJECT

Amazing magnets

Will magnets work through glass? You can answer this question by placing some paper clips in a clear glass jar. Bring a magnet near the outside of the jar. Can you make the paper clips move? Fill the jar with water. Does this change the result of your experiment? Can you get the paper clips out of the water without wetting your fingers?

Try the same experiment using a plastic yogurt container instead of the glass jar. Does the magnet still work through the plastic container?

Next, place the paper clips on a plastic tray. Move the magnet underneath the tray. What happens? Does the same thing happen if you use a metal tray?

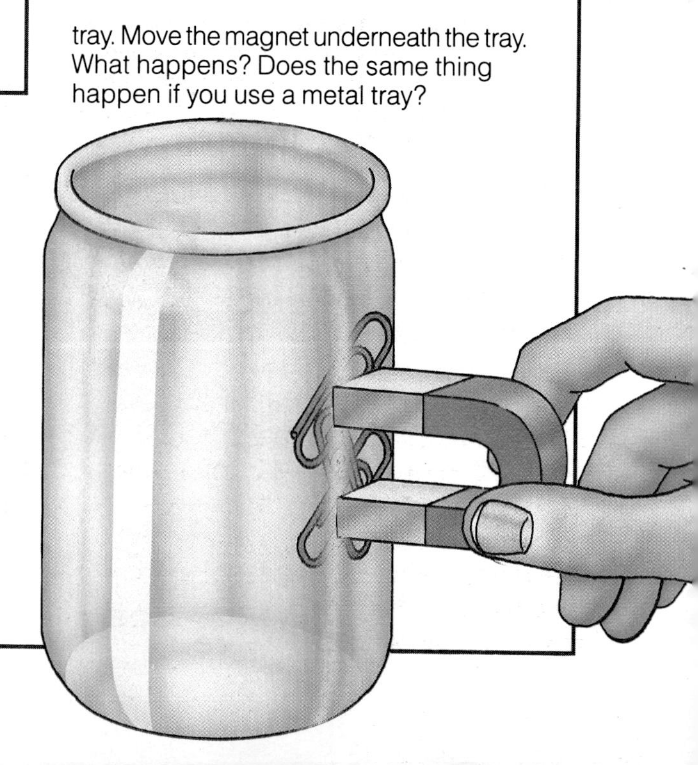

The Aurora Borealis are sheets of glowing light that are seen in countries in the northern hemisphere. They are caused as particles from the sun are trapped by the earth's magnetic field and strike the top layer of the atmosphere, causing it to glow.

dangerous, and would harm the astronauts as well as electrical instruments in the spacecraft.

Some other planets also have magnetic fields. Space probes like *Pioneer* and *Voyager* have discovered them around Mercury, Jupiter, Saturn, and Uranus. The magnetic fields around the giant planet Jupiter are 250,000 times greater than those around the earth. These powerful magnetic fields trap huge clouds of deadly particles which would soon kill any astronaut visiting the planet.

SCIENCE PROJECT

Magnetic fields

You can make a kind of drawing of a magnetic field in the following way. Put a magnet under a piece of paper. Then sprinkle iron filings on the paper. Tap the paper gently, and you will see the filings form a pattern of lines – lines of force.

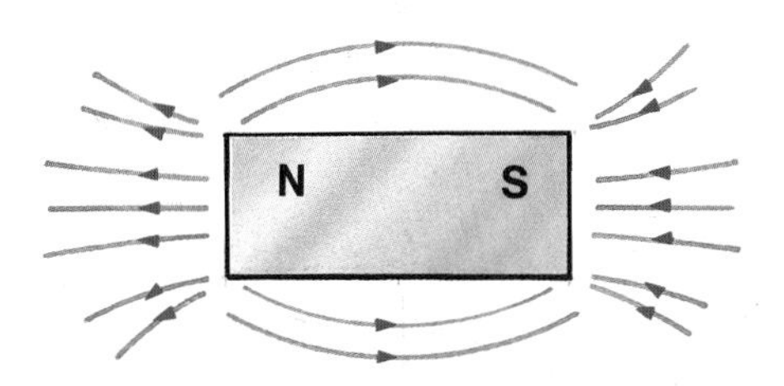

Try the same experiment using two magnets. Place the north pole of one magnet opposite the south pole of the other magnet. Draw a picture of the pattern the iron filings make.

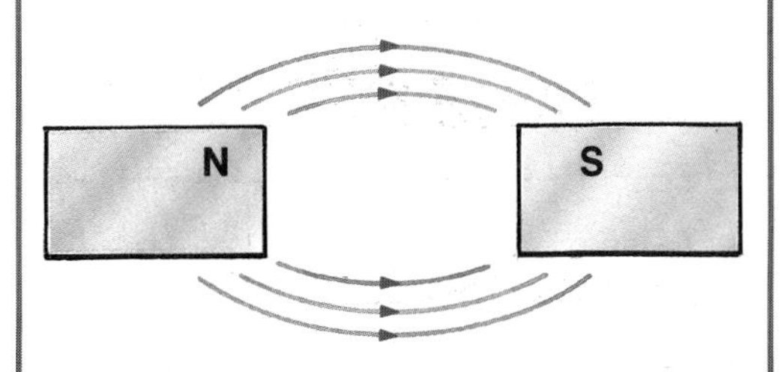

Now place the two south poles of the magnets together. Is the pattern different?

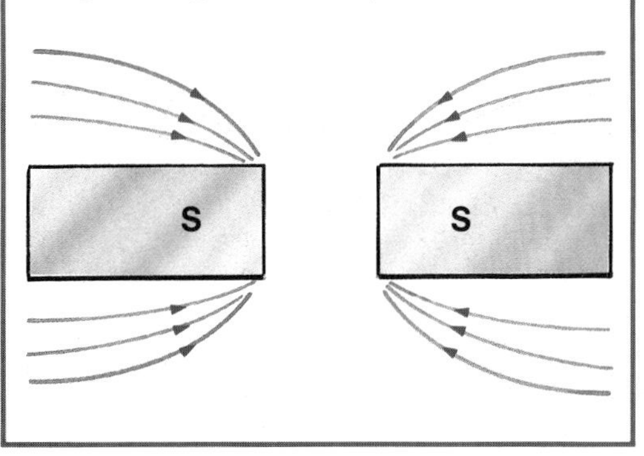

A paper clip chain

Which of your magnets is the strongest? You will need about 20 paper clips and several magnets.

STEP 1 Use a magnet to pick up one paper clip. Place a second clip on the bottom of the first clip. They will cling together because the magnetism is flowing through the first clip to the second one.

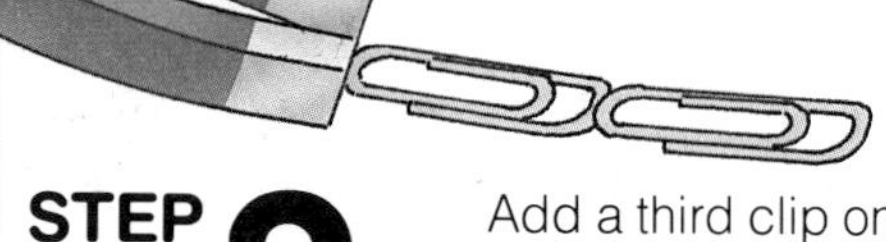

STEP 2 Add a third clip on the bottom of the second. Continue to attach clips. How many clips can the magnet hold in a chain?

STEP 3 Repeat the experiment with other magnets. How many paper clips will each of the different magnets pick up in a chain?

Gone fishing

For this game, you will need some cardboard, paper clips, a piece of string about 12-15 inches (30-40 cm) long, a thin stick about 20 inches (50 cm) long, a shallow plastic tray, and a magnet.

STEP 1 Cut out fish of different shapes and sizes from the cardboard. Color them if you like, and attach a paper clip to each fish.

STEP 2 Tie one end of the string to the magnet and the other end to the stick. This makes your fishing rod.

STEP 3 Put the fish in the tray. Try to pull the fish out using the magnet. How many can you catch?

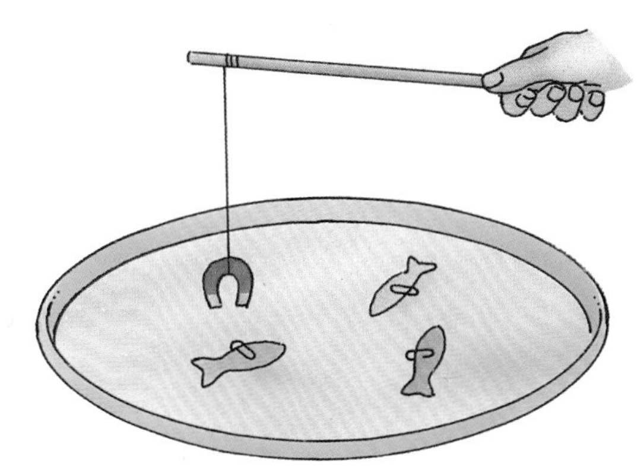

Sailing boats

For this game, you will need a cork, some wooden toothpicks, a small piece of paper, thumb tacks, four blocks of wood, a shallow plastic tray, a magnet, a stick about 20 inches (50 cm) long, and a piece of string about 12-15 inches (30-40 cm) long.

STEP 1 Make a small boat by cutting the cork in half lengthwise. Use a toothpick to make a mast, and a piece of paper for a sail. Push two thumb tacks into the bottom of the boat.

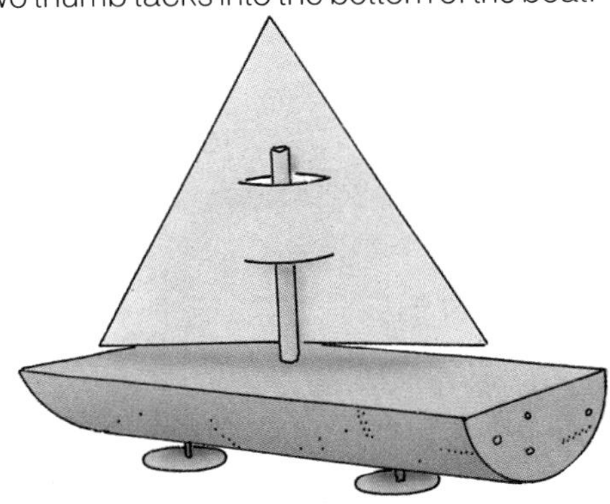

STEP 2 Fill the tray with water. Stand the tray on the four blocks of wood, so there is a small space underneath. Put the boat into the water.

STEP 3 Tie the magnet to the stick with the string, and go sailing! Push the magnet under the tray and use it to move your boat around on the water.

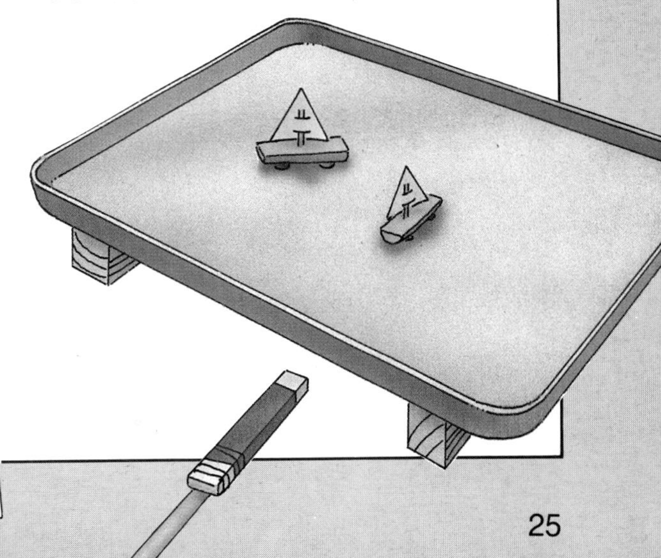

COMBINED FORCES

In 1820, a Danish scientist named Hans Christian Oersted made a discovery. He noticed that a compass needle on his bench moved around when electricity flowed through a nearby wire. This showed that the electric current in the wire was producing magnetism. This discovery was the first hint that magnetism and electricity are connected.

Later, a French physicist named André Ampère wound a wire into a coil. This made a stronger magnetic force when electricity flowed through the wire. Then, he put an iron rod inside a coil of wire. When electricity flowed through the wire, the iron rod became a strong magnet. This kind of magnet became known as an ***electromagnet***.

In 1831, an English scientist named Michael Faraday discovered that when a magnet was moved near a wire, it made electricity flow in the wire. He made a machine for making electricity, called a ***dynamo***. A metal disk was turned between the ends of a horseshoe magnet, which produced a steady electric current. This was the start of the electrical industry, and within 60 or 70 years, power stations were generating electricity using magnets.

Many of the modern gadgets found in our homes today, like telephones and doorbells, use the magnetism produced by an electric current. Others, like washing machines, food mixers, vacuum cleaners, and electric razors, use electromagnetic motors to produce movement.

When the flow of electrons along an electric wire fluctuates very rapidly, ripples of electricity and magnetism are produced. These ripples, called

Michael Faraday was a dedicated scientist who made many important discoveries. As well as inventing the electric motor, generator and transformer, he also studied the effect of electricity on many chemicals.

electromagnetic waves, spread out and travel through space like waves of water over the surface of a lake. Light, radio waves, infrared and ultraviolet radiation, X-rays, and gamma rays are all electromagnetic waves. Radio waves are used to carry signals which can be received by radios and televisions.

IT'S ELECTRIFYING!

Michael Faraday found that if a coil of wire is moved through a magnetic field, an electric current is induced in the wire. He used this principle to build a generator, or dynamo. He also found that if a wire carrying an electric current is placed in a magnetic field, the wire moves: this principle is the basis of all electric motors.

Generators and power stations

Generators can produce a ***direct current***, in which the electrons flow in only one direction, or an ***alternating current***, in which the electrons flow first in one direction and then the other. Faraday made a generator which produced an alternating current by rotating a coil of wire back and forth between the poles of a fixed horseshoe magnet. As the coil turned toward the north pole, a current was induced which flowed in one direction. When the coil turned toward the south pole, the current was reversed, and the electrons flowed in the opposite direction. The current, measured in ***amperes*** using an ***ammeter***, varies from nothing to a maximum as the coil moves in one direction, and then decreases to nothing before building up to a maximum as the coil moves in the opposite direction.

Power stations use this principle to generate electricity. An alternating current is used because, unlike direct current, it is easier to carry over long distances, and its voltage can be

This power station burns coal. Inside the station, the coal is burned to make steam. Some power stations burn oil to make steam. Nuclear power stations use a powerful fuel called uranium. The steam produced is used to run turbines, which rotate the generators that make the electricity.

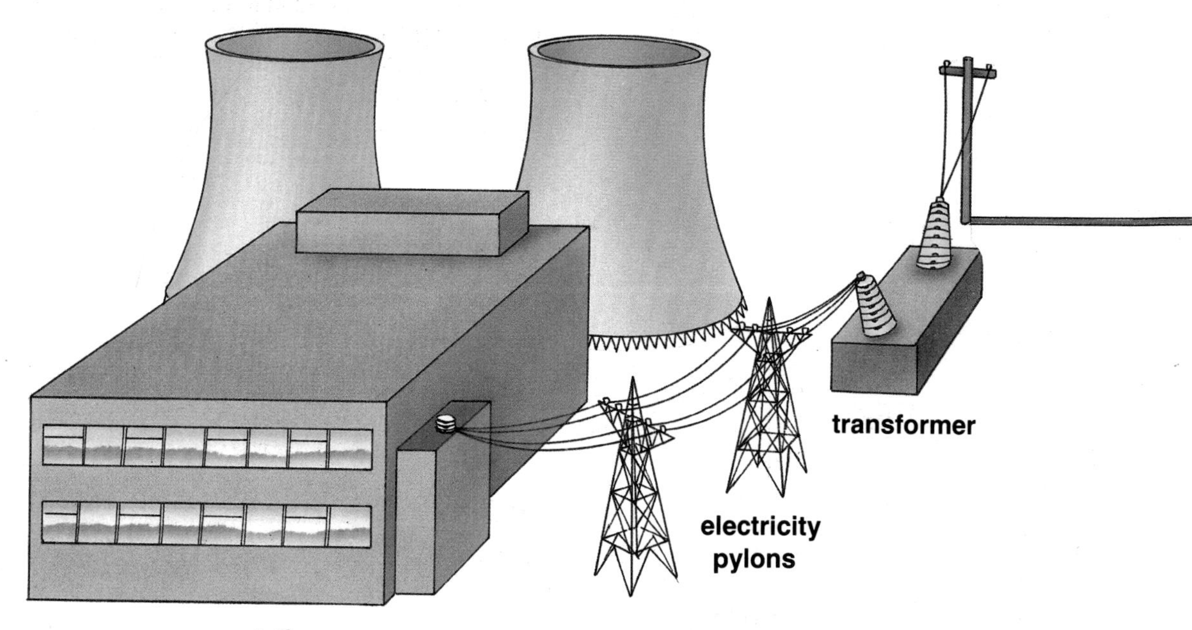

altered by a transformer. This means that the voltage of the electricity supplied to our homes can be lowered to 120 volts, while that supplied to a factory can be a higher voltage if necessary.

From power house to our house

The diagram below shows how the electricity generated in a power station is conducted to your home. From the power station, the electricity is sent along thick wires held up off the ground by pylons. The wires have a very high voltage. Near a town, there is a transformer. Another of Faraday's inventions, transformers are used to increase or decrease the voltage of an alternating current. Lower voltages are safer for use in our homes and offices, so in this case the transformer will make the voltage lower.

From the transformer, thinner wires on wooden poles or inside underground cables are used to carry the electricity to our homes or offices. They eventually enter our houses to provide electricity. There is an electric meter in each house to measure how much electricity is used.

SCIENCE IN ACTION

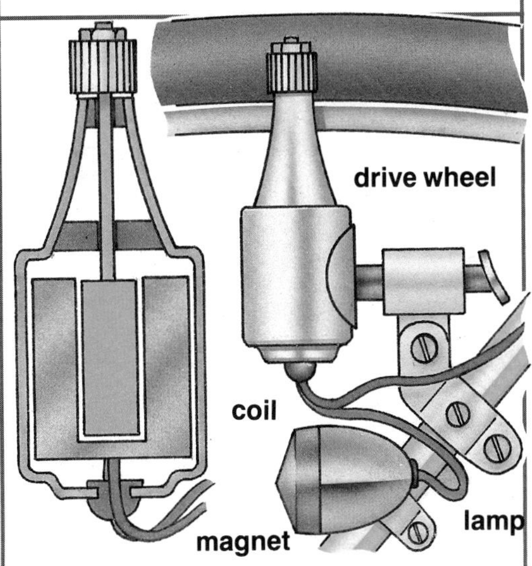

The dynamic dynamo

A bicycle dynamo makes electricity to work the lights on a bicycle. Inside the dynamo, there is a magnet and a coil of wire. The magnet is connected to a drive wheel that rubs against the tire. As the bicycle moves along, the drive wheel turns, which turns the magnet. An electric current is produced in the coil and flows to the front and back lights along a single wire. It returns to the dynamo through the frame of the bicycle.

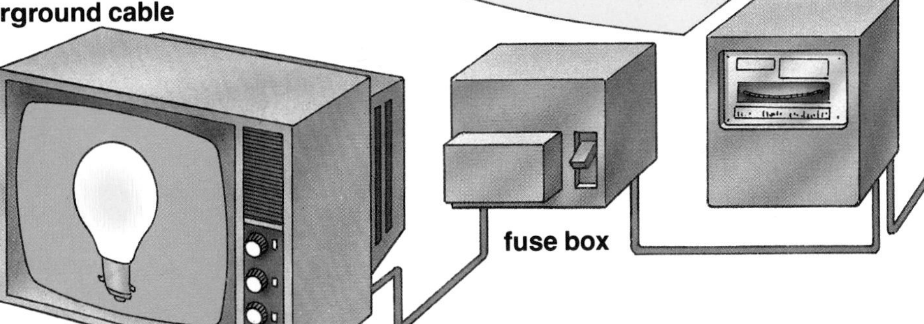

What is electromagnetism? These two projects will help you understand how an electric current can produce a magnetic field.

PROJECT 6

Magnetic currents

Can an electric current produce a magnetic field? You will need a piece of copper wire about 8 inches (20 cm) long, a 4½ volt battery, and a small pocket compass.

STEP 1 Connect one end of the copper wire to one terminal of the battery. Place the wire over the compass so that the wire lies along the compass needle.

STEP 2 Touch the free end of the wire to the other terminal of the battery. Watch what happens to the compass needle as the electric current flows through the wire. Remember, do not keep the battery connected for too long, or it will go flat.

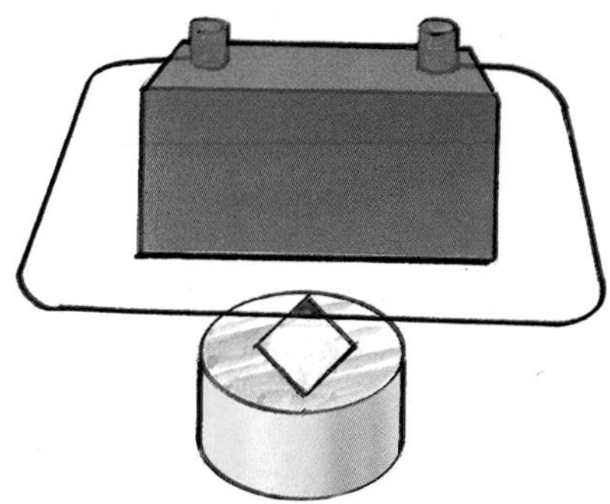

STEP 3 Disconnect the wire from one terminal of the battery. The compass needle will move back to its original position. Place the wire underneath the compass so that the wire lies beneath the compass needle. Reconnect the wire to the battery. What happens to the compass needle? Does it turn in the same or the opposite direction?

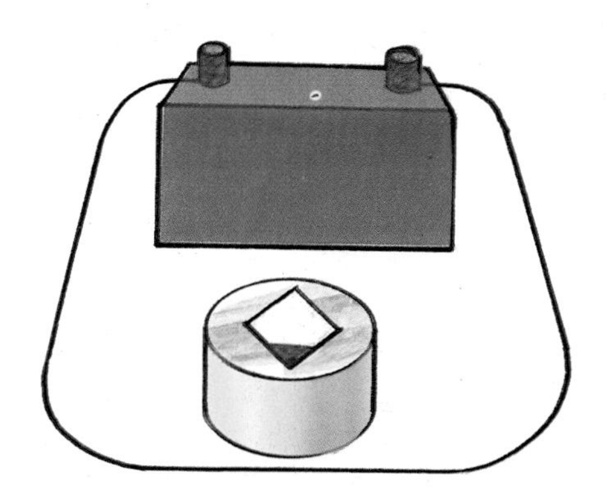

Make your own magnet

You can use electricity to make a magnet. You will need a long length of insulated copper wire, a 4½ volt battery, a pencil, a large steel nail, and a pocket compass.

STEP 1 Wind the wire tightly around a pencil. Remove the pencil. The coil of wire that is left is called a solenoid.

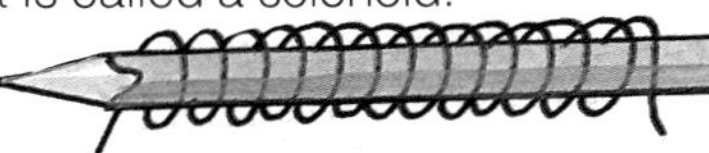

STEP 2 Connect the ends of the wire to the battery. Place the compass near one end of the coil. Watch what happens to the compass needle. Next, put the compass near the other end of the coil. What happens to the needle? Which end of your coil attracts the north pole of the compass and which end attracts the south? The solenoid is acting like a bar magnet – one end is a north pole, and the other end is a south pole. What happens if you connect the wire to the battery the other way around, so that the direction of the current is changed?

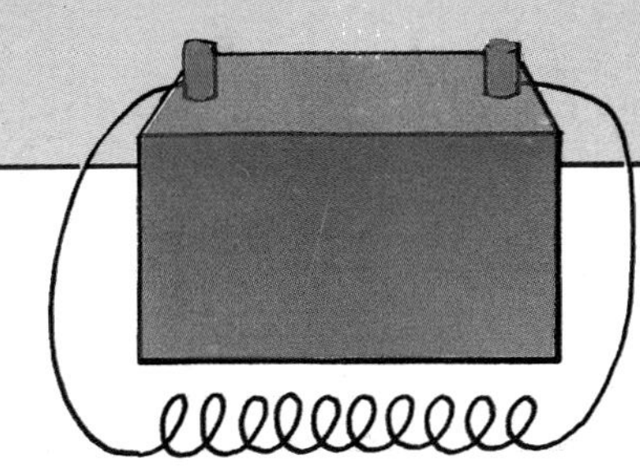

STEP 3 Disconnect your solenoid from the battery and wind it around the nail. This will give you an even stronger electromagnet. Reconnect the ends of the wire to the battery, and repeat the experiment with the compass. Does the compass needle move in the same way?

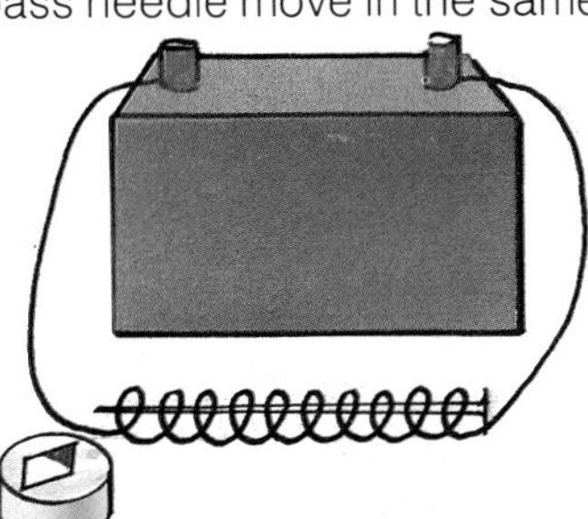

STEP 4 Disconnect one end of the wire and remove the nail. Place one end of the nail near the compass. What happens to the needle? Now place the other end near the compass. Does the nail have a north and a south pole? If so, you have made a permanent magnet.

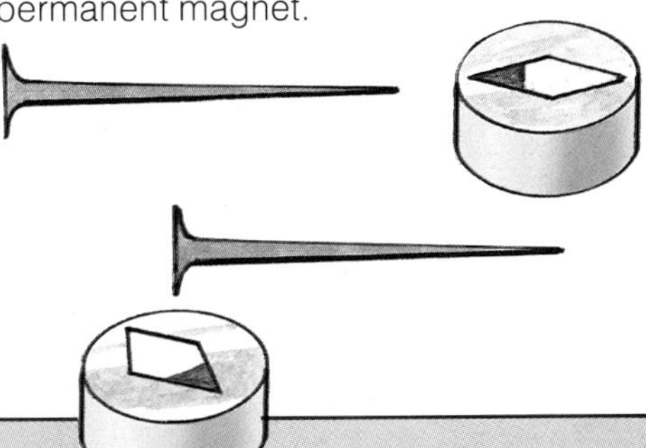

ELECTRIC MOTORS

An electric generator turns movement – provided by water, coal, or oil – into electricity. An electric motor does the reverse. It turns electricity into movement. Electric motors power many kinds of machinery. They range in size from tiny devices that operate an electric toothbrush to huge ones that have the power to run a locomotive. An electric motor changes electrical energy into mechanical power to perform work. Look around your house. How much work is done by electric motors? There are motors in sewing machines, hair dryers, food mixers, washing machines, and so on. Without motors, industry, transport, and the modern home would grind to a halt.

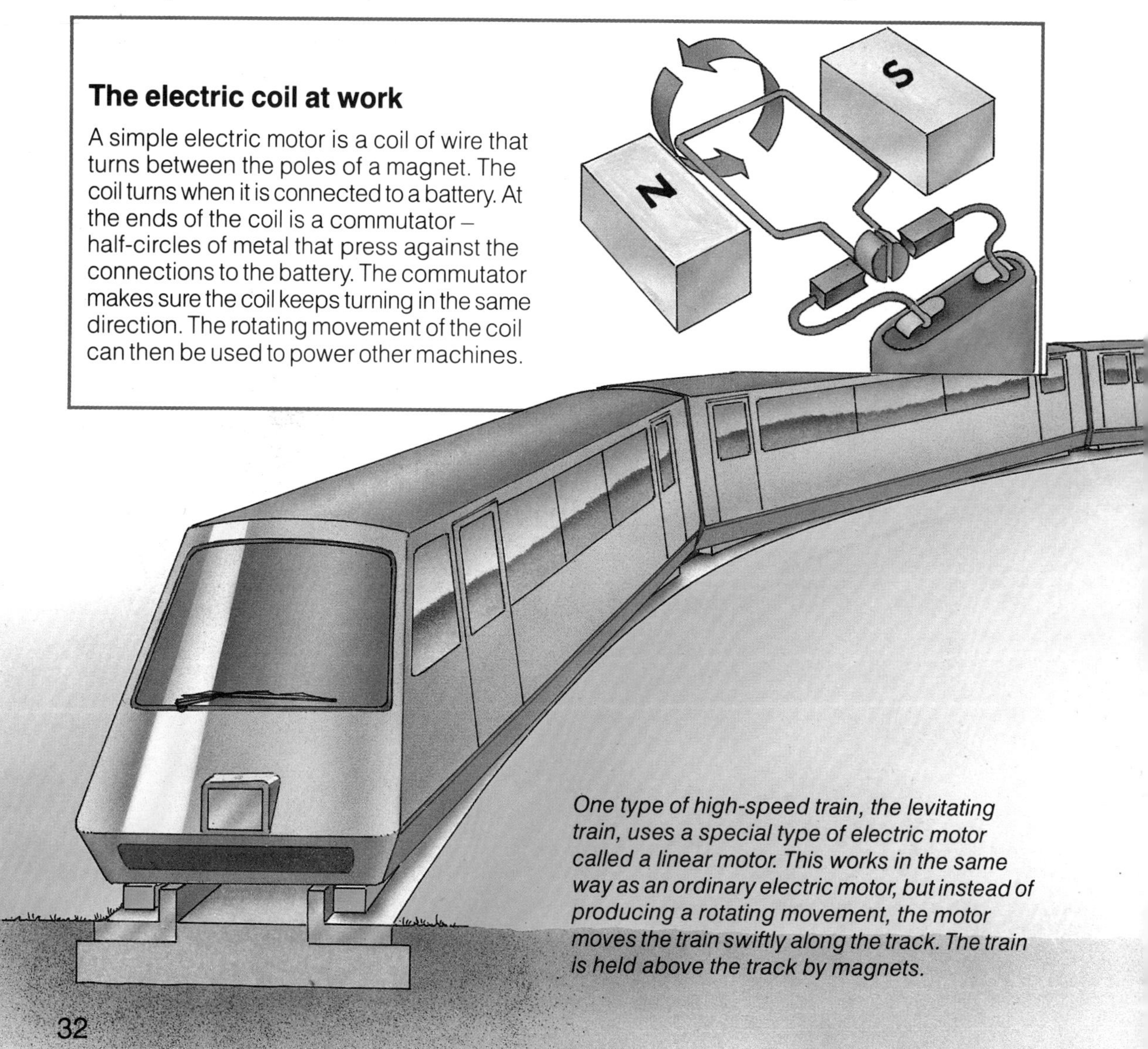

The electric coil at work

A simple electric motor is a coil of wire that turns between the poles of a magnet. The coil turns when it is connected to a battery. At the ends of the coil is a commutator – half-circles of metal that press against the connections to the battery. The commutator makes sure the coil keeps turning in the same direction. The rotating movement of the coil can then be used to power other machines.

One type of high-speed train, the levitating train, uses a special type of electric motor called a linear motor. This works in the same way as an ordinary electric motor, but instead of producing a rotating movement, the motor moves the train swiftly along the track. The train is held above the track by magnets.

The vacuum cleaner uses an electric motor to turn a fan. This sucks air through a pipe into a dust bag, which collects the dust.

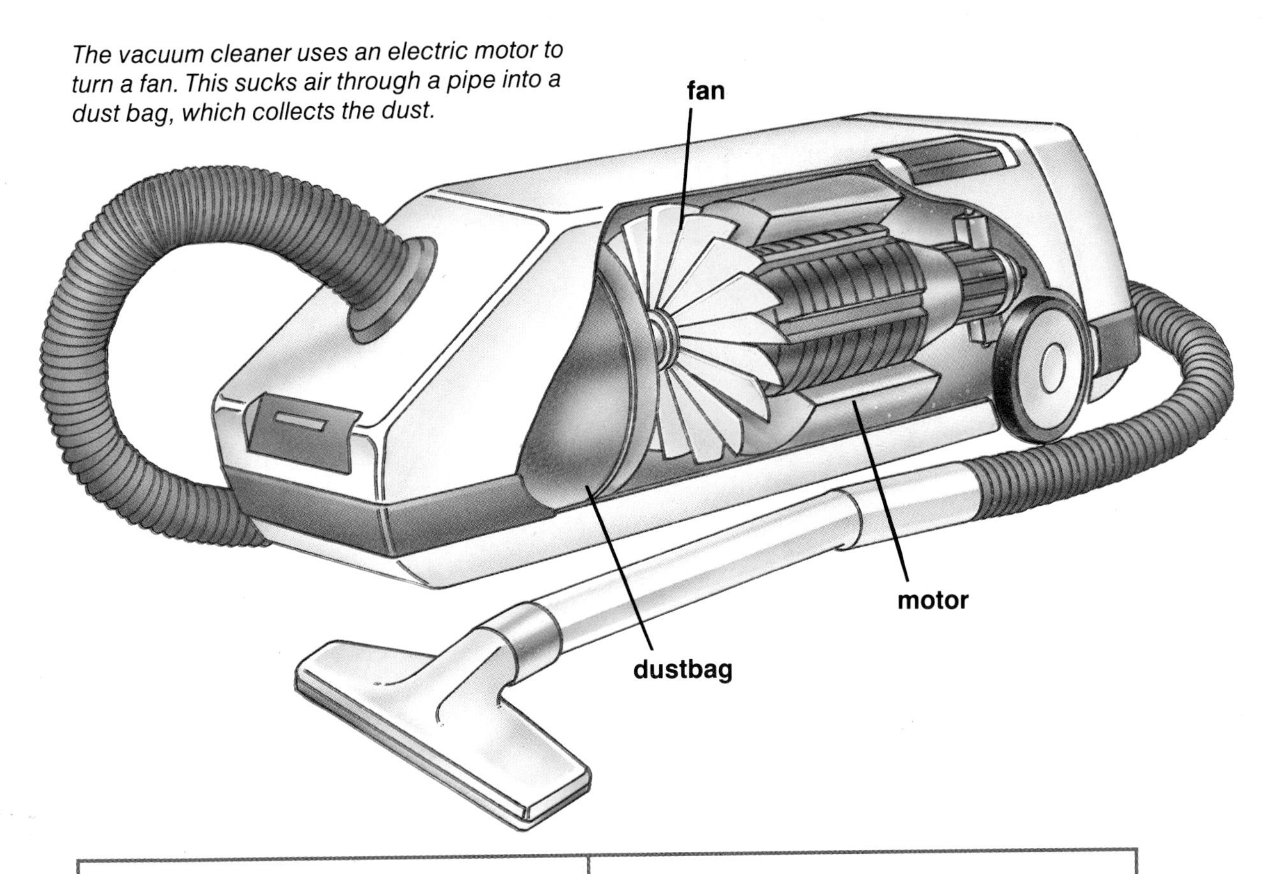

SCIENCE DISCOVERY

Michael Faraday

The three inventions that started the electricity industry were all made by a single scientist – Michael Faraday. He discovered the principles of the electric motor in 1821, and the transformer and electric generator in 1831. Faraday was born in 1791 in London, the son of a blacksmith, and worked for a bookseller, where he read about science. One day, he attended a talk by a famous scientist, Humphry Davy, and wrote down what Davy said in a notebook. He sent the notebook to Davy, who was so impressed that he offered Faraday a job as his assistant. Faraday made many discoveries. In addition to inventing the electric motor, electric generator, and transformer, and discovering alternating current, he discovered how electricity affects different chemicals, and identified an important new substance called benzene.

PROJECT 8

Make an electric motor

You you will need a darning needle 4 inches (10 cm) long, a large cork 1½ inches (3 cm) long, a thimble, 6 pins 1½ inches (3 cm) long, 5 yards (5 m) of thin plastic-covered wire, tape, a piece of thin board like balsa 4 by 6 inches (15 by 10 cm), two paper clips, two thumb tacks, a 6 volt battery, and two horseshoe magnets.

STEP 1

Push the darning needle through the center of the cork lengthwise, wearing a thimble so that you don't stick the needle into your finger. Make sure that the needle goes through the middle of the cork – check this by turning the cork between your fingers.

STEP 2

Stick two pins into one end of the cork. Make sure that the pins are both the same distance from the needle. Leave about ½ inch (1 cm) of each pin sticking out of the cork.

STEP 3

Wind 30 turns of the thin wire around the cork lengthwise, starting and finishing at the end of the cork with the pins in it. Secure the wire with a piece of tape.

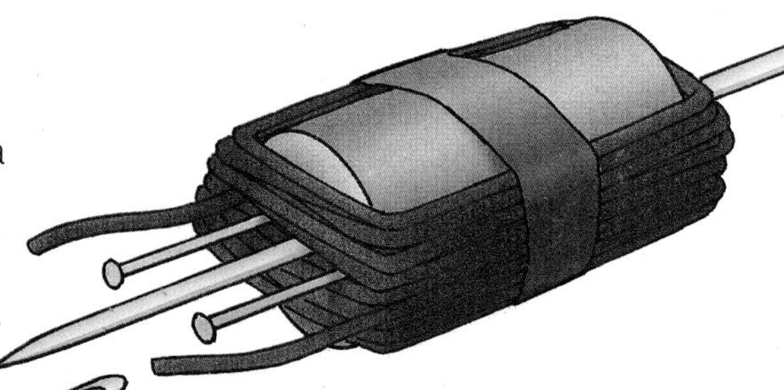

STEP 4

Remove 1 inch (2 cm) of the plastic from each end of the wire. Wrap one of the bare wires around one of the pins, and the other bare wire around the other pin.

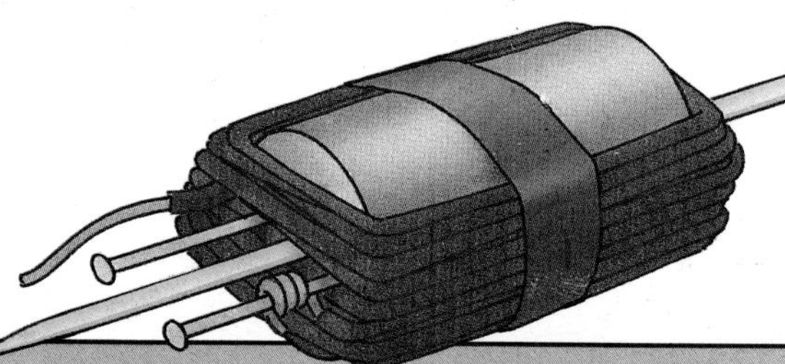

STEP 5

Stick two pins into one end of the wood, so they form an X shape. Do the same at the other end of the wood. Rest the needle on the pins so that the cork is above the board and can turn easily.

STEP 6

Bend a paper clip into an "L" shape. Using a thumb tack, fasten the clip to the board, so that when the cork rotates, the pins on the cork just touch the bent clip. Do the same thing with the second paper clip, placing it on the other side of the darning needle axle. Make sure the cork can rotate freely. The pins should touch the clips gently as they pass.

STEP 7

Connect the battery to the paper clips, using the leftover wire. Place a magnet on each side of the cork. One magnet should have a north pole uppermost, and the other a south pole uppermost. The upper pole should be at the same height and close to the cork. Spin the cork. The motor will begin to turn.

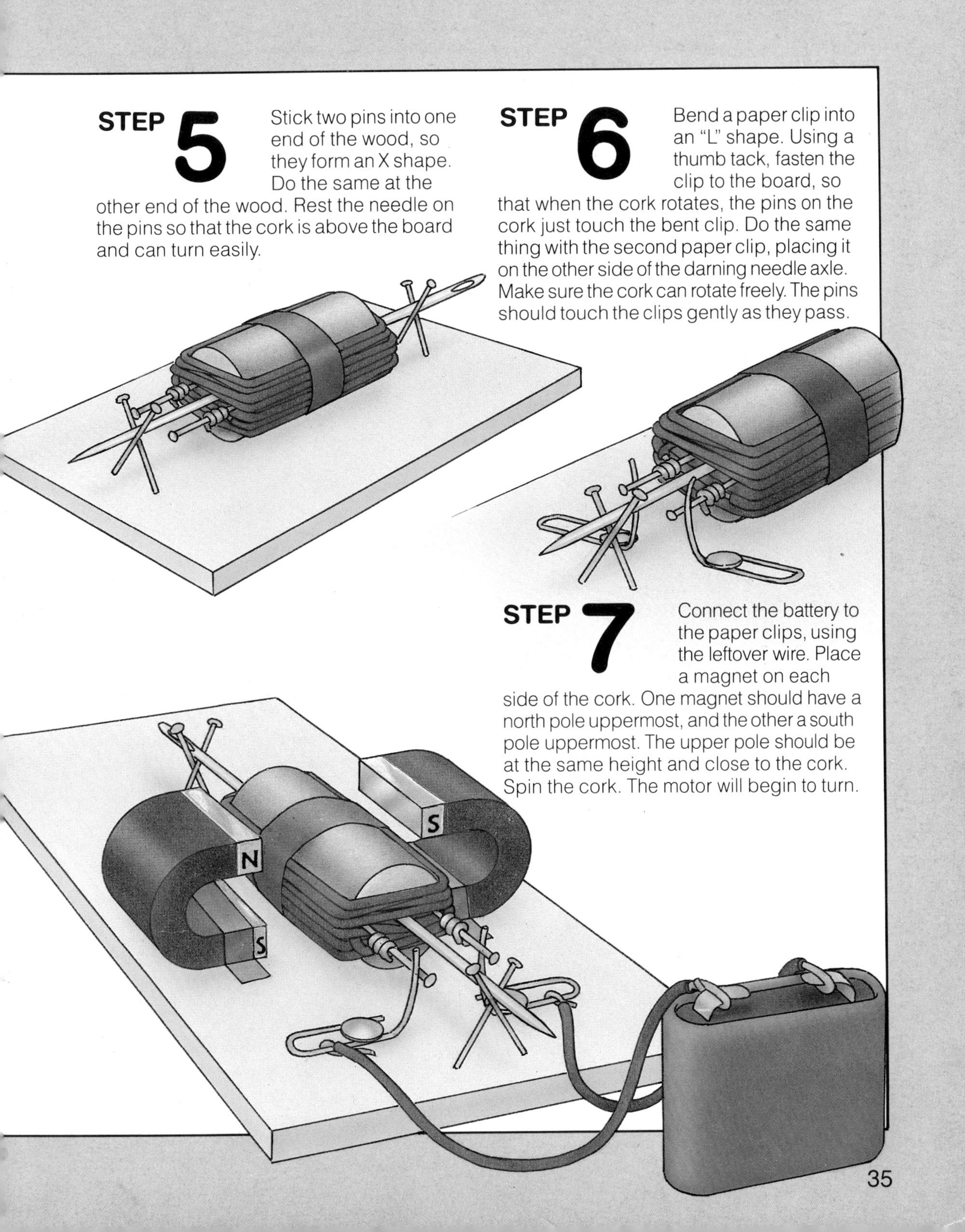

IN THE LABORATORY

Chemistry and electricity are linked, because electricity can change one substance, or chemical, into another. This is a chemical change. Chemistry is the science that studies these changes.

Scientists use electricity to split some chemicals into the simple substances, or elements, that make them up. If an electric current is passed through water, hydrogen and oxygen are produced. This shows that water is made up of hydrogen and oxygen. This splitting process is called ***electrolysis***.

Scientists have also used electricity to discover new elements. Around 1800, Humphry Davy, an English scientist, discovered many new elements by using electricity. His assistant, Michael Faraday, also used electricity in his chemistry experiments.

Today, electricity is used to make many useful chemicals, such as chlorine, which is used to kill germs in swimming pools. An electric current is passed through salt water, and chlorine gas bubbles off. Many metals are made in this way, too, including aluminum, magnesium, and zinc. Electricity is also used to give a hard or shiny coating to a metal object by a process called electroplating. It is used to cover steel knives and forks with silver, for example.

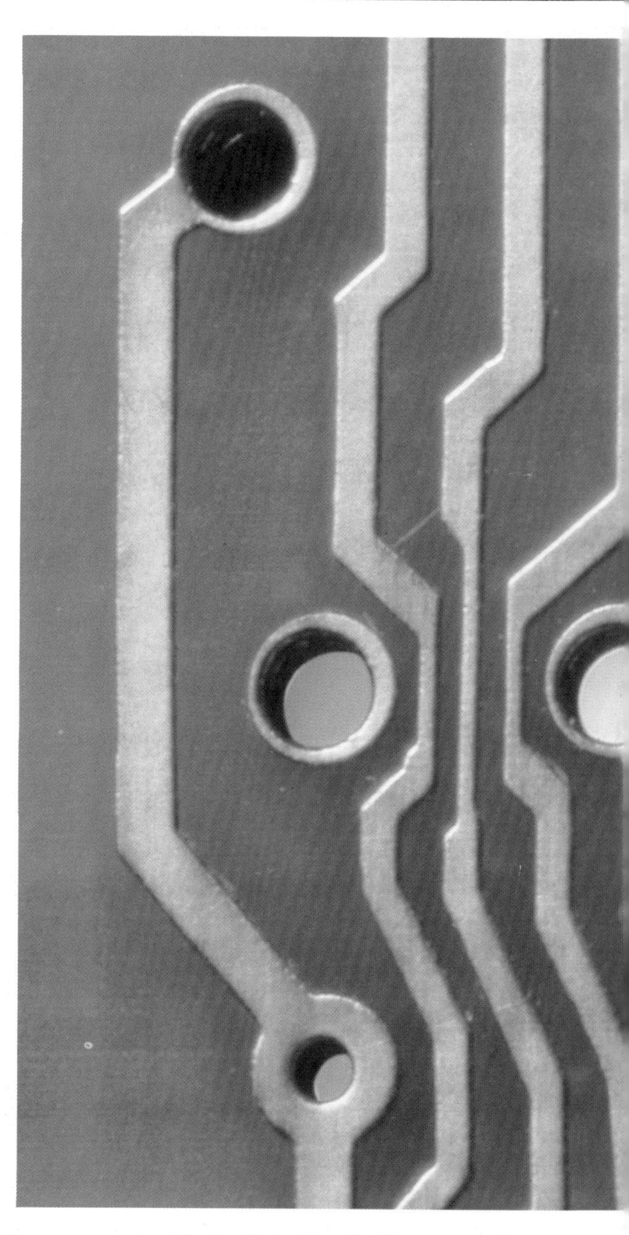

Many batteries use chemical changes to make electricity. This is the opposite of electrolysis. Substances inside the battery react together and produce electricity. When all of the substances have reacted together, the battery stops making electricity and has to be thrown away. There are some batteries, called accumulators, that can store electricity

Electroplating shows how electricity can produce a chemical change. It is used to coat metal objects with a different metal. Printed circuits which have to be of a very high quality often have gold plated onto the etched copper links.

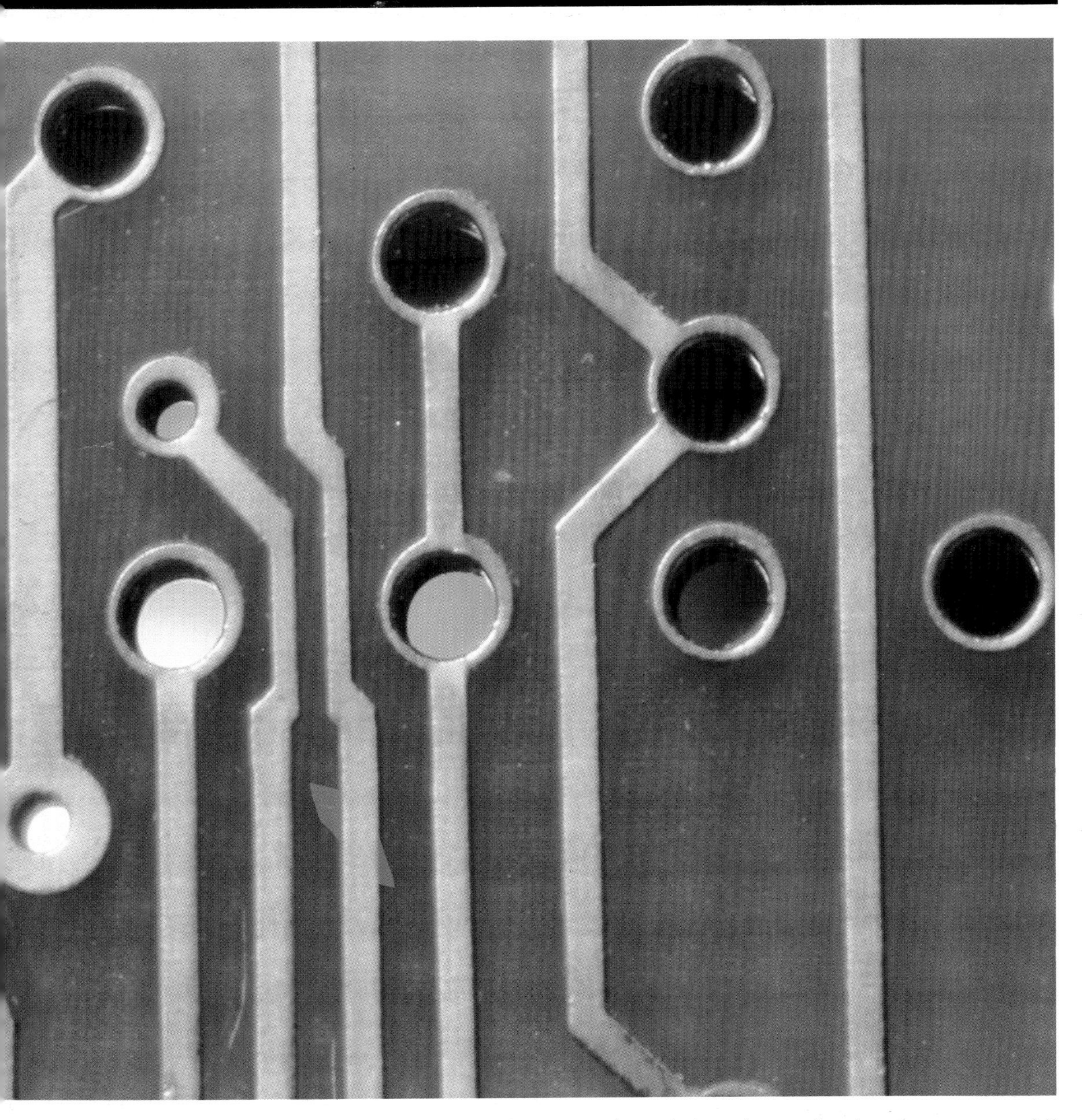

and can be refilled or recharged when they run down. These batteries are used in cars. There are other types of batteries, called fuel cells, that make electricity from hydrogen gas. All these different kinds of batteries work because of the connection between chemistry and electricity.

POWER PACKS!

In 1780, the Italian scientist Luigi Galvani made a strange discovery. He found that a dead frog's legs would twitch when a spark of electricity passed through them. But the real surprise was that the legs twitched when they were touched with a knife blade. They did not need to be connected to the electricity supply to do this. One day, he took the frog's legs outside while he thought about the mystery, and hung them over an iron fence. Once again, they twitched. Galvani thought this must be because electricity was in the frog's legs, and that all animal flesh must store electricity.

The first battery

Another Italian, Alessandro Volta, a professor at Pavia University, thought that Galvani was wrong. He believed the answer was that the metal knife and iron fence were producing electricity.

In 1800, Volta found that electricity was made when two different metals

A car battery is sometimes called an accumulator because it can store, or accumulate, electricity. When it runs out of electricity, it can be recharged. Inside the accumulator, there are lead plates standing in weak sulfuric acid. As the battery is used, the sulfuric acid becomes stronger and is used up. Distilled water has to be added to keep the battery full and dilute the acid.

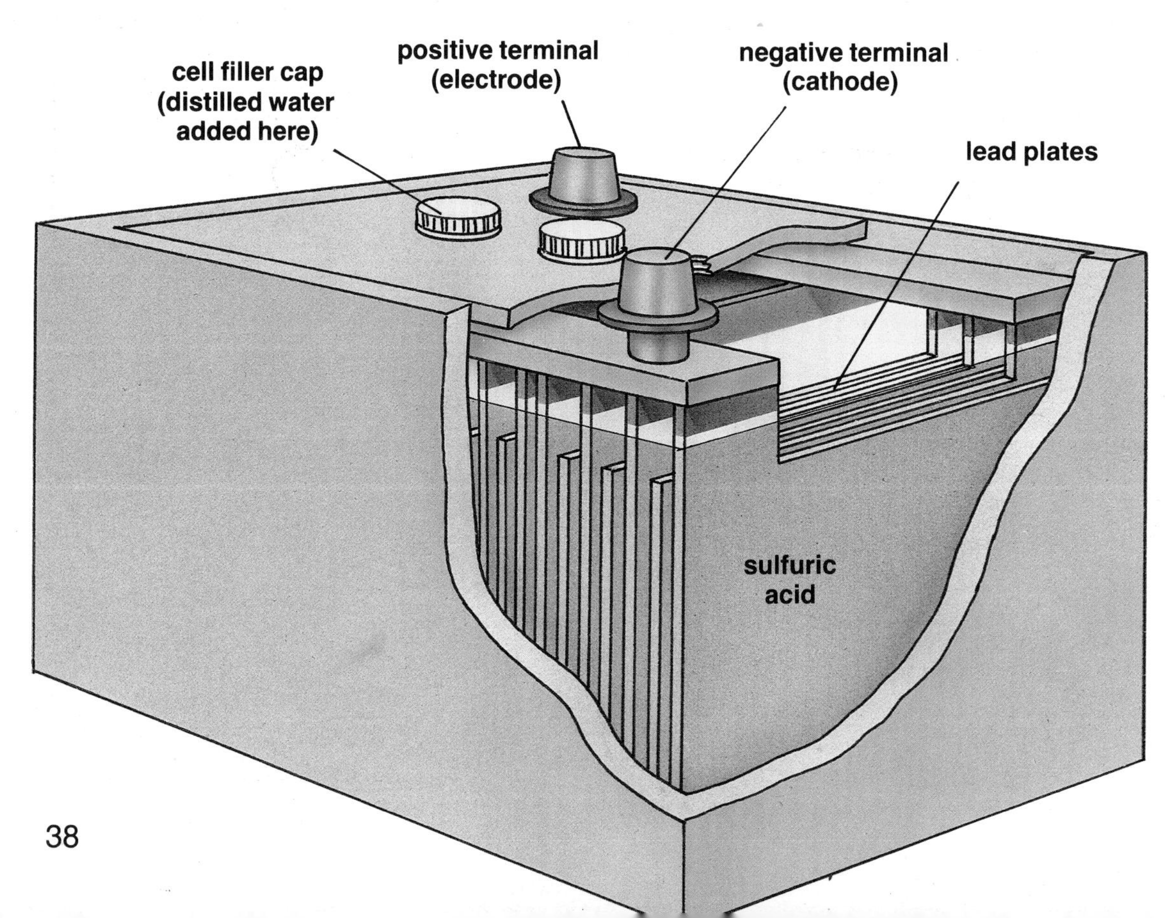

SCIENCE PROJECT

Make an electric cell

You can make a simple electric cell using a lemon and thin strips of copper and zinc, or some copper and zinc nails. First, press the lemon gently on a table to make it juicy inside. Then, stick the two metal strips into the lemon, making sure they do not touch. Touch the two pieces of metal with your tongue, and you will feel a tingling sensation. This is because the metal strips are making a small electric current that is passing through your tongue.

Try using a potato instead of a lemon. Does this make electricity, too?

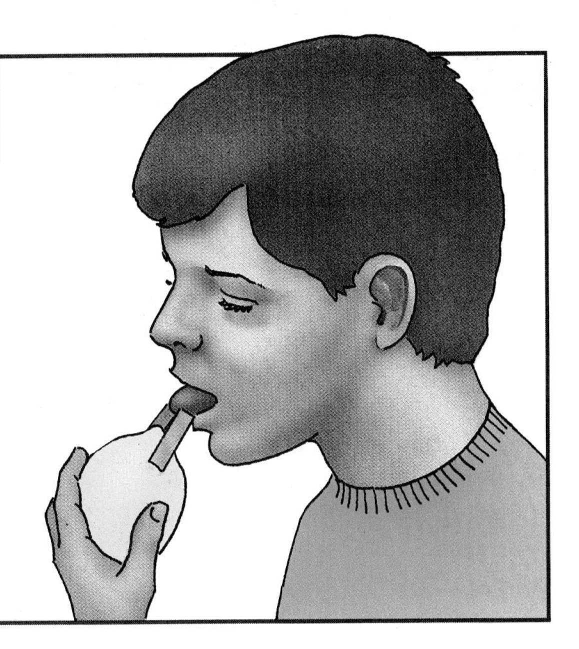

were placed in salty water. There was no need for a frog's, or any other animal's, legs. Volta proved his idea by making some "electric *cells*", which were disks of moist material, like wood, set between disks of copper and zinc. When wires were connected to the metal disks, electricity flowed along the wire. This was the first steady flow, or current, of electricity. Volta made the first electric "battery" by connecting many of these cells together to produce a large current.

Postive and negative electrodes

In a modern electric cell, there is a liquid called *electrolyte*. There are also two rods or sheets, usually of different metals, called *electrodes*. The positive electrode is called the *anode*, and the negative one is called the *cathode*. In the simple lemon cell, the zinc is the cathode and the copper is the anode. The electrolyte is the acidic juice inside the lemon.

A voltaic cell

Volta improved his original electric cell. He used copper as the positive electrode, or anode. The other electrode, called the cathode, was made of zinc. The electrodes were placed in weak sulfuric acid.

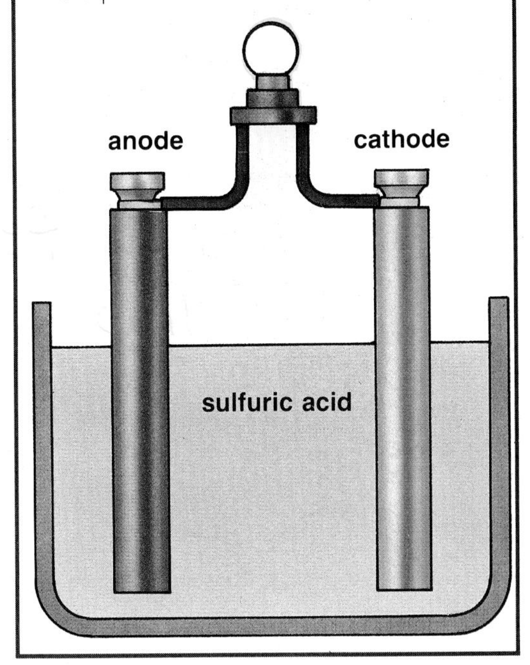

Electrifying changes!

See how electricity can break up, or decompose, some chemicals. You will need a small glass, salt and water, ink, a saucer, some copper wire, and a 4½ volt battery.

STEP 1

Dissolve as much salt as you can in half a glass of water. Add ink to make the water dark blue. Put a little of the inky salt water into the bottom of the saucer.

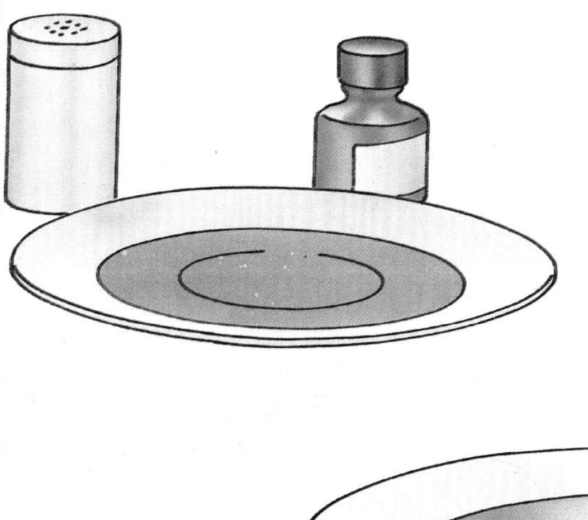

STEP 2

Attach a piece of wire to each terminal of the battery. Dip the ends of the wires into the liquid in the saucer. You will immediately see bubbles coming from the wire attached to the negative terminal of the battery. These are bubbles of hydrogen gas, which are being produced as the electricity decomposes the water.

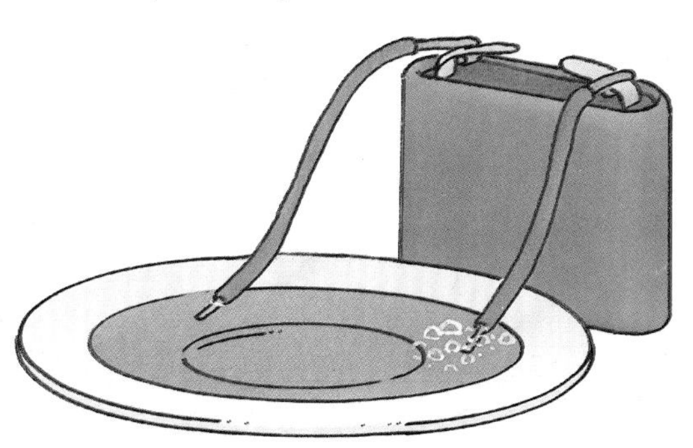

STEP 3

Watch closely and you will see the color of the ink change near the other wire. If you sniff near the wire, you may smell chlorine. This is the gas used to kill germs in swimming pools, and it is being produced as the electricity breaks up the salt in the water. Chlorine bleaches, or removes, the color from the ink. With some inks, a pink color may be seen as this happens.

PROJECT 10

Copper plating

You will need about ½ ounce (10 g) copper sulfate (from a drugstore), a small glass, water, some copper wire, a coin containing copper, a 4½ volt battery, and a large paper clip.

STEP 1

Dissolve the copper sulfate in a glassful of water. Get two pieces of copper wire about 8 inches (20 cm) long. Wind the end of one piece around the copper coin, and attach the other end to the anode, or the terminal marked + on the battery. Wrap one end of the other wire around the paper clip. Attach the other end to the cathode, or the terminal marked – on the battery.

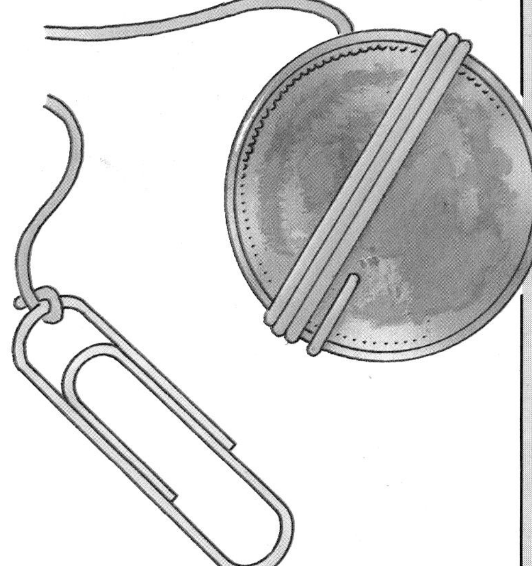

STEP 2

Put the coin and the paper clip into the glassful of copper sulfate and water. Wait. Soon you will see that the coin looks cleaner. This is because copper is being removed from the surface of the coin. At the same time, the paper clip will become covered with a reddish layer of copper.

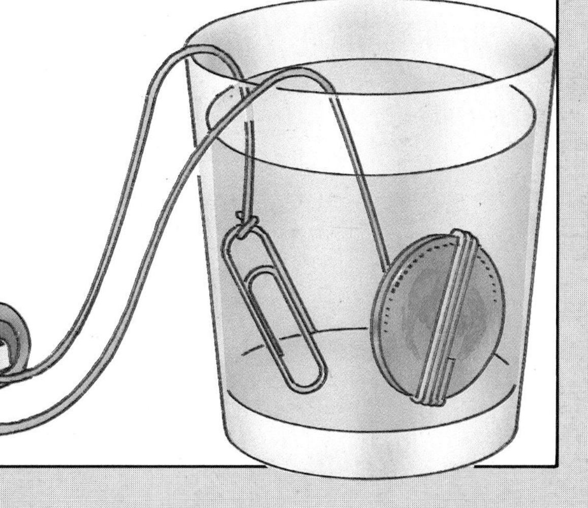

WHY DOES...?

How does an electric doorbell work?

An electric doorbell contains an electromagnet. When you press the bell button, an electric current flows through the electromagnet coils. This attracts the armature, and the striker is pulled onto the bell, which rings. The movement of the striker moves the contact apart, which stops the flow of the electric current. The striker then springs back to its first position. The whole sequence repeats again, and again, and again.

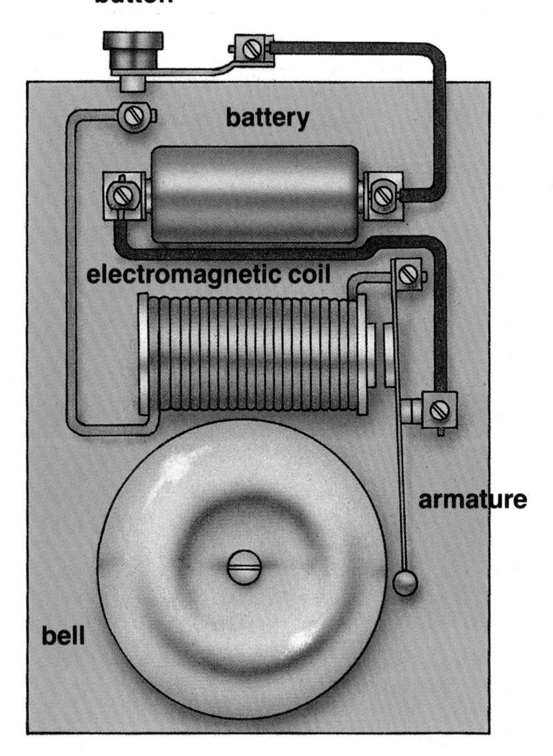

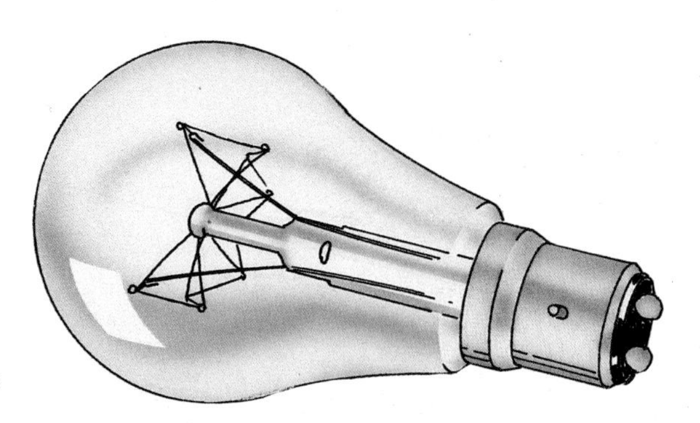

Why are light bulbs labeled 40W, 60W, or 100W?

Thomas Alva Edison, the American inventor, was born in 1847 and died in 1931. One of his numerous inventions was the light bulb. On an electric light bulb, you can usually see a number like 40W, 60W, or 100W, which shows how much light the bulb produces. The higher the number, the brighter the light. The letter "W" stands for "watt," the unit used to measure electrical power, named after James Watt, a Scottish inventor.

On an electricity meter, you might also see kWh (kilowatt-hour); 1 kWh is the amount of electrical energy that a 100 watt bulb would use in 10 hours. This amount of energy would keep a television going for 5 hours, a vacuum cleaner for 2 hours, a one-bar electric fire for 1 hour, or boil 10 quarts (10 liters) of water.

Why are fuses needed in a house?

A fuse is a safety device. It is a piece of wire which melts easily if it is heated. There are fuses in a box near the electric meter in a house. All the electricity used passes through the fuses. There are two kinds of home fuses. Cartridge fuses are long and narrow. They can carry high currents. Plug fuses screw into a socket. They carry lower currents.

How does a telephone work?

The telephone works by changing sound into electricity and back again. The mouthpiece contains a thin metal plate, or diaphragm, that vibrates like the eardrum in your ear. Behind the diaphragm are carbon granules. They are squeezed together or left loosely packed, depending on the pressure of the vibrations. The electric current flowing through the granules becomes weaker or stronger in response to the movement of the granules. The electric current travels by wire to the earpiece of the receiving telephone, which contains an electromagnet and another diaphragm, which vibrates like the vocal cords in your throat. The electric current flows through the electromagnet causing the diaphragm – which is held in place by a permanent magnet – to vibrate and reproduce exactly the sounds received by the mouthpiece.

Why is the earth a huge magnet?

Scientists have found that there is liquid, or molten, rock deep inside the earth, in the core. The core is covered by a layer of solid rock. When the earth spins each day, different parts spin at different speeds. The outer layers spin a little faster than the core. This acts like a giant dynamo, just like that on a bicycle. The spinning of the bicycle wheel produces electricity within the dynamo. In the same way, as the earth spins around, it generates electricity within the core. This produces the earth's magnetic field.

What are semiconductors and superconductors?

Semiconductors, such as silicon, can usually conduct electricity only a little. They are midway between good conductors like metals and insulators like rubber. They are very useful, because if they are heated, or if a small amount of another material is added to them, the amount and direction of the electric current they conduct can be controlled. They are used to make transistors and integrated circuits, which are used in most modern electronic systems.

Superconductors are the best conductors. They have no resistance; once an electric current starts flowing in a superconductor, it keeps going. But a superconductor needs to be kept very cold, often at about 269°C below zero, to work. Scientists have found new substances that do not need to be quite as cold as this, but making wire from them is still very expensive and difficult.

THINGS TO REMEMBER

What the words mean....

Here are some explanations of some of the words in this book that you may find unfamiliar. In some cases, they aren't the exact scientific definitions, because many of them are very complicated. But the descriptions should help you to understand what the words mean.

ALTERNATING CURRENT An electric current that flows in one direction and then in the other direction.

AMMETER An instrument that measures the size of an electric current in amperes.

AMPERE The unit for measuring the size of an electric current, named after a French scientist, André Marie Ampère.

ANODE The positive terminal of a battery or cell.

BATTERY One or more electric cells which make or store electricity.

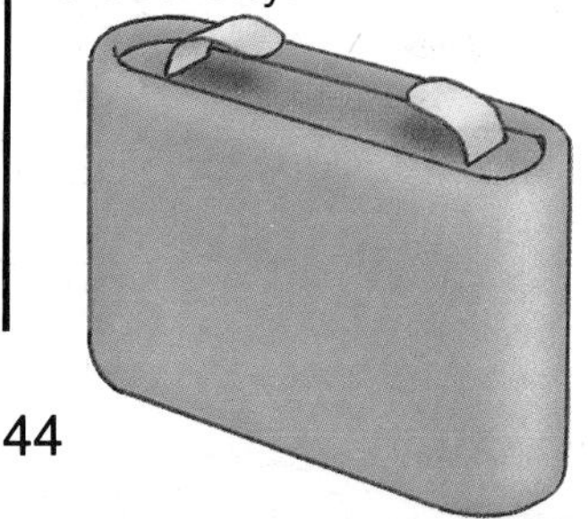

CATHODE The negative terminal of an electric cell or battery.

CELL A part of a circuit that produces, or stores, electricity by using chemical changes.

CIRCUIT The complete path along which an electric current flows.

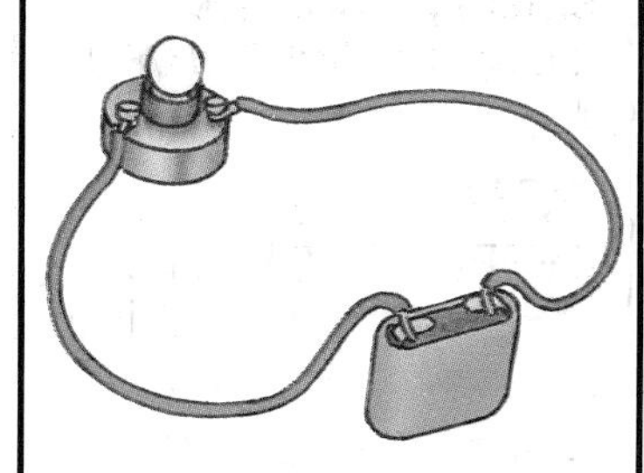

CONDUCTOR A material that allows an electric current to flow easily through it.

DIRECT CURRENT An electric current that flows in one direction only.

DYNAMO A machine that produces electricity from movement.

ELECTRIC CHARGE The amount of electricity held by an object.

ELECTRIC CURRENT A flow of electrons through a wire or other conductor.

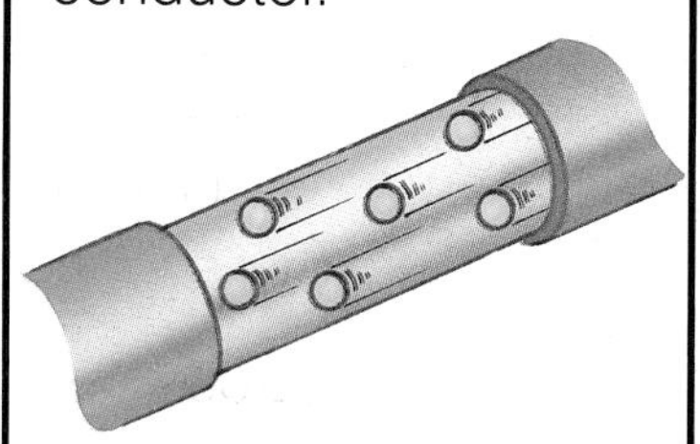

ELECTRODES The rods that carry current into and out of a battery.

ELECTROLYSIS The breaking down of a substance using electricity.

ELECTROLYTE The liquid in a battery or cell.

ELECTROMAGNET An iron bar with coils of wire around it. It acts as a magnet when an electric current flows through the wire.

ELECTROMOTIVE FORCE The electrical pressure or force that drives electrons around an electric circuit.

ELECTRON A tiny particle that has a negative electric charge.

INSULATOR A material that does not let electricity flow through it.

LINE OF FORCE Lines around a magnet that show the magnetic effect. They can be seen by spreading fine iron dust around a magnet.

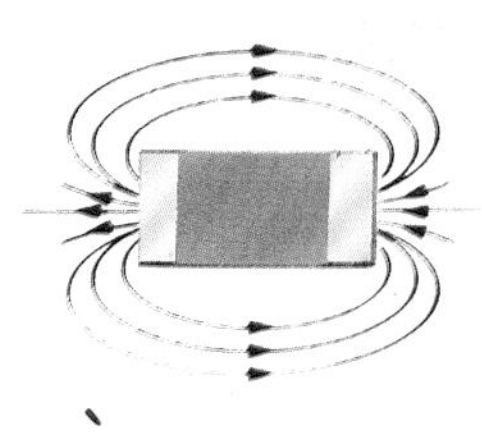

MAGNET A piece of steel or iron that can attract iron or steel. A magnet will attract or repel other magnets.

MAGNETIC FIELD The space around a magnet where the magnetic effects can be felt.

MAGNETIC POLE The place on a magnet where the magnetic effect is strongest.

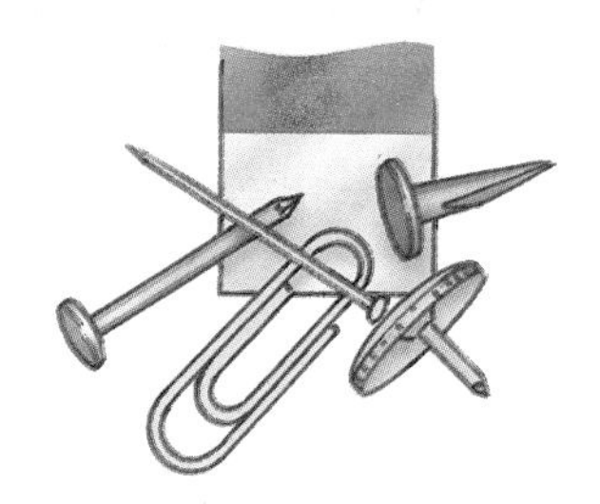

OHM The unit used to measure electrical resistance. The unit is named after a German scientist, Georg Ohm.

PARALLEL CIRCUIT An electric circuit in which the parts of the circuit are connected side by side. The current splits between the parts.

POTENTIAL DIFFERENCE The difference of electrical pressure between points in an electrical circuit. It is measured in volts.

SERIES CIRCUIT An electrical circuit in which the parts of the circuit are connected end-to-end. The current flows through all the parts.

STATIC ELECTRICITY Electricity that is produced when certain materials are rubbed together.

TRANSFORMER A device used to increase or decrease voltage. A transformer only works with an alternating current.

VOLT A unit used to measure the electrical pressure in a circuit, named after an Italian scientist, Alessandro Volta.

VOLTMETER An instrument used to measure electrical pressure in volts.

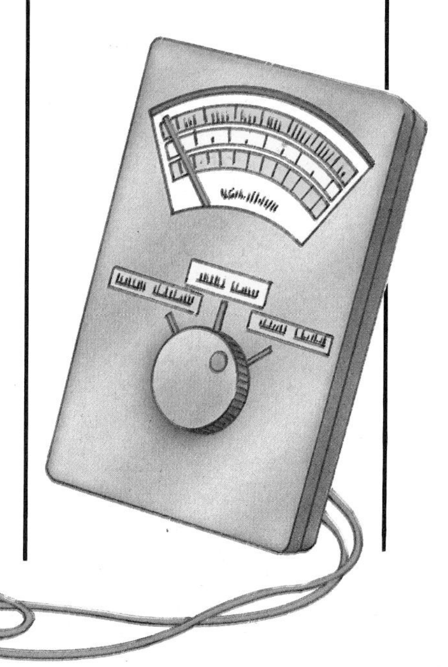

INDEX